Moral
Calculations

Moral Calculations

Game Theory, Logic, and Human Frailty

László Mérő

Translated by Anna C. Gösi-Greguss
English Version Edited by David Kramer

COPERNICUS
An Imprint of Springer-Verlag

Illustrations by Miklós M. Miltényi.

Published in the United States by Copernicus, an imprint of Springer-Verlag New York, Inc.

Copernicus
Springer-Verlag New York, Inc.
175 Fifth Avenue
New York, NY 10010
USA

Library of Congress Cataloging-in-Publication Data
Mérő, László.
    Moral calculations : game theory, logic, and human frailty /
László Mérő.
        p.   cm.
    "Copernicus imprint."
    Includes bibliographical references and index.
    ISBN 0-387-98419-4 (hardcover : alk. paper)
    1. Game theory.   2. Thought and thinking.   I. Title.
QA269.M457   1998
519.3—DC21   98-17443

Manufactured in the United States of America.
Printed on acid-free paper.

9 8 7 6 5 4 3 2 1

ISBN 0-387-98419-4   SPIN 10659411

Preface

This book is about rational thinking—something that may, perhaps, not even exist. There are many indicators that human thinking is essentially nonrational, even in those cases where the methods of pure logic can effectively be applied. I addressed this matter in my book *Ways of Thinking*. The present book does not rely on *Ways of Thinking*, but rather complements it: Here we are going to discuss thinking from the point of view of John von Neumann's theory of games.

I shall not attempt to characterize the mathematical level of this book, because it has none. Here you will find no formulas whatsoever, although you will occasionally meet ideas whose origins are mathematical. To understand the book you will need no mathematics beyond the four basic arithmetical operations. Nevertheless, I do not promise the reader an easy time of it, and I defend myself with the words of Frigyes Karinthy, the father of Hungarian satire: "It's not me that's complicated; what's complicated is what I'm talking about."

This book is in the form of an essay. It is intended primarily to be read rather than studied, although parts of it could be taught in a university course. Therefore, I did not stick to the fundamental rules of textbook writing, namely, that the material be easy to learn, easy to teach, and easy to examine on—usually at the expense of being slightly boring. Rather, my aim was to meet the demands of the reader seeking mental adventure. Nevertheless, I found it useful to differentiate between generally accepted scientific theories and facts on the one hand, and my own views on the other. For the former I use the first person plural or passive voice, while for the latter I employ the first person singular. Although even what is presented as science should not be viewed as indisputable, I definitely hope that my opinions will provoke debate and further thought.

The book is divided into three parts. The first part introduces the basic concepts of game theory, together with several games and real-life situations that illustrate the theory's ideas and typical approaches to different psychological and moral issues. The second part is probably the most difficult of the three. Here I demonstrate in five different fields of science—mathematics, psychology, evolutionary biology, economics, and quantum physics—the diversity of ways in which the ideas of game theory can manifest themselves. In the third part the lines converge. This part is purely about psychology—about rational thinking in particular—which is not at all the same as logical thinking.

I would like to thank Csaba Andor, László Antal, Éva Bányai, Miklós Barabás, Nóra Bede, Anikó Bódi, Judit Bokor, Ferenc Bródy, Zsuzsa Csányi, István Czigler, Andrea Dúll, Péter Futó, Éva Gartner, Csilla Greguss, János Herczeg, Mónika Holcsa, Györgyi Hosszú, Sándor Illyés, András Joó, Zsuzsa Káldy, Ildikó Király, Erika Kovács, Éva Kovácsházy, Kriszta Mády, Csaba Mérő, Katalin Mérő, Vera Mérő, Nóra Nádasdy, Balázs Nagy, János Pataki, Ferenc Pintér, Júlia Sebő, István Siklósi, Endre Somos, Dóra Speer, Eszter Szabó, Judit Szabó, Péter Tátray, Enikő Tegyi, Róbert Urbán, Zoltán Ülkei, Tibor Vámos, Katalin Varga, Zoltán Vassy, and Zsuzsa Votisky for their ideas, remarks, suggestions, and criticism. The preparation of this book was generously supported by grant No. T–006845 from OTKA to the author.

László Mérő
Budapest, Hungary

Contents

GAMES OF MORALITY

1

Auction and Posing

Boxing is a sport in which the winner is also badly beaten.

The Dollar Auction Game

In a game described by Martin Shubik a dollar bill is put up for sale. It is offered at auction with a minimum bid of one cent. Anybody who bids this much can take the dollar, provided, of course, that no one else offers more. The game proceeds according to the usual rules of auctions, but with one exception. The special rule is that the auctioneer must be paid not only by the highest bidder, but also by the second-highest bidder. The highest bidder pays what he bid and takes the dollar, while the second-highest bidder pays what *he* bid but gets nothing.

Shubik first published an account of this game in 1971, noting that in his experience at social gatherings the dollar bill sold on average for $3.40. Shubik's profit was greater than this, since he also received the runner-up bid, and so he raked in almost seven dollars. This game has subsequently been used in several well-designed psychological experiments, with very similar results.

The rules of this game may appear absurd: Why should the player who was outbid have to pay? After all, he didn't buy anything. Well, be that as it may. Absurd or not, mature, intelligent adults entered into the game—of their own free will—and were willing to pay three to four dollars for a dollar. Not, mind you, for some object valued at a dollar, whose subjective value might be anything, but for an ordinary, legal tender, one-dollar bill.

The Three Critical Points

A sober mind finds it difficult to understand such behavior. Yet such behavior does occur. Shubik wrote: "A large crowd is desirable. Furthermore, experience has indicated that the best time is during a party when spirits are high and the propensity to calculate does not settle in until at least two bids have been made." It is worth limiting the bids to increments of ten cents, to prevent someone from spoiling the game by bidding 99 cents immediately, which would make further bids pointless, since after such a bid no one else could make a profit. (Even in such cases somebody occasionally decides to punish the spoiler, offering 100 cents for the dollar in the hope that the spoiler will lose his 99 cents. Thus the game can get going in this way, too, although it starts off at a higher level.) Usually, however, there are three critical points in the course of the game.

The first critical point is whether the game begins at all. At lively parties it almost always does. It is enough for the auctioneer to propose the game, explain the rules, and play the clown a little: "Well, doesn't anybody want to buy a dollar for one cent? All right, I have a bid of one cent. Doesn't anybody want to put in his two cents to win a dollar?"

Once the first two bids have been made, the machinery proceeds under its own power. During bidding one player may think, "My 20 cents has just been outbid. Shall I quit now, when I could still get the dollar for 22 cents?" But alas, his opponent is thinking along the same lines: "I'd rather pay 23 cents for a dollar than lose 21 cents."

The second critical moment occurs when the bids reach the 50-cent mark. At this point the next player must bid at least 51 cents, and it may well occur to him that if the bidding continues, the auctioneer will make a profit. But he usually brushes off these dark thoughts by telling himself that nonetheless *he* can still get a good bargain. At this point it may help if the auctioneer urges the group along a bit, but usually this is unnecessary. Once the bidding has reached 50 cents, it will almost certainly continue up to 99 cents.

The third critical point occurs when someone offers 100 cents for the dollar. At this point the bidder may still be under the illusion that he can escape without a loss. But his opponent knows that if he drops out of the bidding, he will be out 99 cents, while by bidding $1.01 he loses only 1 cent. He knows that the bidding has become irrational and he will now end up in the hole, and moreover, the auctioneer will at least double his investment. He hopes that his opponent will finally come to his senses and drop out, so that he will lose only 1 cent instead of 99. But after the 101-cent bid the opponent finds himself in a completely analogous situation: He will lose a dollar if he stops, but only two cents if he continues. After this, the bidding generally escalates rapidly—to the great amusement of the onlookers. The two bidders, on the other hand, are *not* amused—once a married couple competed in the auction and wound up going home in separate cabs. On another occasion the "winner" paid twenty dollars for the dollar, while his opponent stopped bidding only because he had run out of cash.

Out of the Party and into the Psych Lab

Scientific psychological experiments are not carried out at parties but in sparely furnished laboratories, with perfectly sober subjects, under exactly reproducible conditions. Nevertheless, such experiments have led to results similar to those of Shubik's social games.

Here, too, the experimenter exercised mild pressure on the players at the beginning, which he did by repeating a fixed text, to ensure that the behavior of the subjects did not depend on the experimenter's improvisation. If the first two bids were given, everything continued as at Shubik's parties.

Such experiments raise several methodological and ethical issues. While it may be acceptable at a party to win a few dollars from one's friends, it is improper to fleece experimental subjects. It is also improper to incite friends against each other. Although the psychologist may pacify the grumbling souls at the end of the experiment and give back their money, the subjects must not know this at the outset, since otherwise they would not take the game sufficiently seriously and thus would not behave in the laboratory as they would in real life. We will not go into detail as to how scientists have avoided such problems, but their consistent results in various experiments, all very similar to Shubik's results in social situations, lead us to venture some far-reaching conclusions.

At the beginning of the game, both at Shubik's parties and in the formal experiments, several players often entered the bidding; but in the end, the auction was always reduced to two bidders. The greater the number of participants, the greater the chance for the game to get started; if there were more than ten, a first bid was almost invariably made. In one series of experiments, forty groups of college students were studied, and in every case the students went over the one-dollar limit; in fact, in half the cases bidding ended only when one of the players offered all the money he had on him, whereupon his opponent simply outbid him.

Subjects who entered the bidding spiral displayed intense emotion. They perspired; they stared in distress; sometimes they shouted. In one experiment psychologists measured the subjects' galvanic skin response, heart rate, and other measures of stress. When the subjects passed the one-dollar limit, they generally evinced changes characteristic of extreme tension—similar to what parachutists experience just before jumping out of an airplane—such as a sudden decrease in pulse rate.

In discussions after the experiment most bidders reported that their opponents had gone completely crazy: Is it normal to bid more than one dollar for a dollar? Very few players admitted that they themselves had done the same. Several students who had participated in a previous session but had not bid, later entered the game and bid more than one dollar—despite having seen the outcome in the previous group. Subsequently, they said it had never occurred to them that the same thing could happen to them.

Even in experiments with only two subjects, the bidding went over one dollar almost half the time, and subjects frequently battled it out to their last cent. The situation was similar if the auction was not for cash, but for some object. In such cases, the moment when a bid went over the object's subjective value (as perceived by the subjects) proved to be a particularly critical point. There was now no turning back (as for a parachutist at the moment of jumping). From this point on there was no bound on the object's worth. This dramatic phenomenon, appearing in every experiment, was named the *Macbeth effect*, in reference to Macbeth's lines, "I am in blood / Stepped in so far that, should I wade no more / Returning were as tedious as go o'er."

In the experiments with two subjects, players seldom learned anything from their mistakes. In one experiment, for example, every subject had to participate in two games. Sometimes a person playing the game for the first time was paired with a subject who in his first trial had bid over one dollar, and sometimes two "experienced" subjects were paired. In both cases, few of the previously embarrassed players avoided the trap the second time.

In the dollar auction game a greater proportion of men than women bid over a dollar. Before jumping to premature conclusions, however, let me mention that we shall see a game in Chapter 3 where women were more likely to fall into an equally sticky trap. We do not yet know the reason for this gender difference. The phenomenon is statistically significant, but it is far less exact than a chromosome test: With games there is a slight, but definite, difference between the sexes.

In one of the experiments, the game was interrupted before each bid, and the players were required to answer a short questionnaire. These opportunities for reflection did not have the expected sobering effect. As usual, the bids usually rose above a dollar, and often much higher.

However, the answers to the questions posed to the subjects changed considerably in the course of the bidding. At first, most players stated that they were bidding primarily in order to win money. As the bidding progressed, the importance of monetary gain gradually diminished, giving way to competitive feelings: "I'll show who's better." "I won't be made a fool of."

High-Stakes Auctions

When researchers analyzed President Johnson's speeches on the Vietnam War given between 1964 and 1968, they found that his rhetoric changed dramatically as the war escalated. At first the President spoke of democracy, freedom, and justice. Later, his speeches were dominated by concepts like national honor, halting the spread of Communism, and avoiding the appearance of weakness. This transformation is strangely similar to the changes in motivation expressed by subjects in the dollar auction game.

Shubik's first article on the game appeared in 1971, when the Vietnam War had reached its most bitter and hopeless stage. Shubik introduced the game as a model of the increasingly meaningless escalation, although he later reported that this was not what had led him to invent the game. He was not even sure whether the game was his invention or whether it was a brainchild hatched together with some of his colleagues. At the time he was looking for a theoretical framework, in the form of a simple and abstract game, in which to study the mechanisms of addiction.

In science, a good question is often more valuable than ten good answers. Shubik's original question led to profounder results than he had expected. It was soon realized that the phenomenon exhibited in the dollar auction game is very general. Its essence is strikingly phrased in the title of a book by A.I. Teger: *Too Much Invested to Quit*.

This phenomenon is also called the Concorde fallacy. The costs of the Concorde, the supersonic airplane developed jointly by the British and the French, rose steeply during its development, and it soon became clear, at a time when only a small part of the original development budget had been spent, that the enterprise would never bring a profit. Nonetheless, both the British and French governments became more and more entangled in the project, until in the end several times the original development budget had been spent. It would even have been more cost-effective to terminate the whole enterprise just before tightening the last screw. The Concorde has continuously operated at a loss, but it had become a prestige investment, and the British and French are still proud of it. Here, too, the psychological phenomena of the dollar auction game are in operation.

The example of the Concorde exactly matches the logic of the dollar auction game with respect to one of the players (the investor). For the other player (the designer), raising the costs, that is, continuing the bidding, is a perfectly rational strategy. This player partly takes over the role of the auctioneer—for example, it is in his interest to keep the opening bid as low as possible. The investment, however, progresses essentially as in the dollar auction. The psychological mechanisms of escalation can be modeled by the dollar auction game in almost its pure form.

Dollar Auctions in Everyday Life

I have told many of my friends about the dollar auction game, and they all have said that they certainly would never get involved in such a harebrained enterprise. I do not think my acquaintances are any wiser than Shubik's: Many of them would certainly have gotten entangled in the bidding. In fact, we all have found ourselves in similar circumstances many times. Indeed, several everyday situations operate according to the logic of the dollar auction.

The longer we wait for a bus to come, the more difficult it is to decide to hail a taxi, even if before heading for the bus stop we had seriously considered taking a cab because we were in a hurry. The longer we watch a terrible movie, the more likely we'll watch it to the end, although it is increasingly less likely that something interesting will happen. Television programmers know this very well, so they schedule many commercials toward the end of a film, when viewers are less likely to switch to another channel during the break.

Strikes also follow the logic of the dollar auction. Often, the damage done by the strike would exceed the cost of meeting the strikers' demands, while the workers' lost wages are much greater than what the satisfaction of their demands would bring in decades. Nevertheless, each side tries to hold on a little longer than the other, because the loser will get no compensation for the damage caused by the strike, or for the lost wages. In long strikes the debate visibly shifts from financial matters to matters of principle—just as do the values of the players in the dollar auction. Toward the end of the strike, an agreement on financial matters could be reached quite easily, but by that time the struggle is no longer about money. In these cases a skilled negotiator can be of invaluable assistance. It is a time-honored practice for the negotiator to introduce a new item for negotiation, one that neither the employers nor the employees had thought of before. For example, at an appropriate moment he brings up the maintenance of work uniforms. If the parties can agree to this after a short debate, then they can end the strike with neither side losing face.

Competitive bidding also obeys the logic of the dollar auction. All of the participants invest time and money to prepare their materials for the competition. The greater the investment in preparation, the greater the chances of winning the competition. Yet only one applicant will win. The others will have worked for nothing.

The principle of the dollar auction keeps many people in unsatisfying jobs and unhappy marriages. Every fight is basically a dollar auction; if it isn't, it is not a fight but a thrashing.

Dollar Auctions in the Wild

Certain animals compete over mates or territory not by fighting, but by standing face to face in a threatening posture, ready to fight, and remaining that way for a long time. Finally, one of them backs down, and the other wins the valued commodity. This method of solving conflicts is common among animals living in strict hierarchical societies, but it also occurs in animals that do not live in groups, rarely meet one another, and have little memory of the outcome of previous fights. This form of combat is particularly common in well-protected animals for whom injury is very unlikely. For them fighting does not make much sense, because the outcome would depend greatly on chance. Furthermore, an injury could be fatal. For similar reasons, animals with strong offensive weapons may also settle their disputes by some kind of posing.

Animals that do not live in groups pay for such posing fights in the currency of time. No matter how valuable the commodity in question, no animal can afford to spend too much time posing—they have other vital things to do. The great tufted titmouse, for example, when feeding its young has to find prey every half minute; each second of the day is valuable. Thus, instead of posing, they settle their differences by open combat—which is more dangerous for both competitors.

No matter which animal wins the posing fight, the price of the commodity is paid by both. They have frittered away their time in posing. Thus, posing fights follow *exactly* the rules of the dollar auction. The player "bidding" the last second wins the goods. The other player gets nothing, but he pays essentially the same as what the winner paid.

At first the dollar auction may have seemed rather artificial, but in fact, we have seen that in nature, matters sometimes proceed exactly this way. We should no longer be surprised at how general and fruitful a model this game has proved to be.

The two animals entering a posing fight could easily settle their dispute by tossing a coin, instead of by tiring and time-consuming posing. There is not even an auctioneer who might prevent them from coming to an agreement. Still, this solution is forbidden by natural law. Not because sticklebacks cannot toss a coin. (Oh, yes. Sticklebacks, who also settle disputes by posing, are the favorite little fish of ethologists.) If the survival of the sticklebacks depended on whether or not they could equitably draw lots among themselves, natural selection would long ago have developed a species of sticklebacks capable of drawing lots. But fighting is the means of natural selection. The individual that shows its fitness in combat will get the goods. The fight must be difficult and self-sacrificing, even if it does not lead to physical injury.

Let us assume for a moment that each individual can somehow exactly calculate the value, in posing time, of the desired commodity. The animal considers every important aspect: the role of the object in its survival, the animal's own condition, how much lost time it can make up, and many other things. The animal thereby determines how much time it can afford to spend on posing. It is not worth posing for a longer time; the animal is not a human being—who can afford such irrationality. (Why is it that we humans can afford such a luxury? That is the main theme of this book.)

If two equal opponents stop posing exactly when it is no longer worth it, both of them will leave the situation at the same moment and will get nothing for their pains. Thus it may be worth posing a little longer. But in the long run this is not worthwhile either, since if the individuals of a species always entered into games ending with a net deficit, the species would soon die out. There seems to be a stalemate: One should not pose for less time than the value of the goods, because that is a certain loss; but neither should one pose for more time, nor for the same amount of time.

Posing for a Random Period

Mathematicians have an artful suggestion for sticklebacks to help them resolve this stalemate. They (the sticklebacks, not the mathematicians) should spend on posing a more or less random amount of time. Before each fight, each party should choose a range of times it would be willing to spend posing, and then choose at random a value within this range. This way *every individual will continue posing for a period that cannot be predicted beforehand.* If my opponent backs out within that period, I win, but if my predetermined time passes, I immediately give up. Thus, while the exact period cannot be predicted before any fight, the random generator should be calibrated so that on average it reflects the real value of the goods. A businessman would doubtless add some profit, too, but biological fights are purely for survival.

An advantage of this strategy is that it permits the fighting parties to avoid the trap of the dollar auction. If the object of the fight—say a female—is worth five minutes of posing, than each individual will pose for a predetermined period of time; it may pose for only 3 or 4 minutes, or as much as 6 to 8, or perhaps exactly 5.

If the animals use this procedure, an interesting equilibrium may develop, determined by a mathematical formula describing the probability with which we should choose a certain duration. If every individual of a species behaves this way, then the given species will gain a selective advantage over every other species that settles its posing fights some other way—all else being equal, of course. Therefore, once a species begins to use the suggested strategy, all rival species must switch to this mode of behavior in order not to suffer a selective disadvantage. If the competing species poses longer than the prize is worth, even its winners will lose in the long run. If they pose for a shorter period, they will rarely win, while if they always pose for exactly as much time as the commodity is worth, their behavior will be too easy to predict, and they will lose. Furthermore, if the competing species chooses a random strategy that is different from our optimal one, it will once again be at a disadvantage.

We have so far assumed that the goods in question are worth the same to every individual. Generally, this is not the case: For a strong individual a given commodity may be worth a longer posing time, because the animal will become less tired and can make up the lost time more easily. It possesses a selective advantage. In the case of a random posing strategy, the stronger individual can afford, on average, a longer posing period.

Other factors may influence the selected duration of posing. A female may be worth more to an older male who senses that this is his last chance of begetting offspring. The same territory may be worth more to someone who is already living there than to an intruder.

At the beginning of the fight, the parties may not know exactly how much the given good is worth to the other. According to the strategy based on chance, they do not have to know this: It is sufficient if each knows its own valuation and chooses its random posing duration accordingly. An equilibrium still develops.

Do sticklebacks know about all this? Do animals really carry on their posing disputes according to this strategy, or are this formula and the balance that arises mathematical minutiae alien to life as we know it? We are not asking how a stickleback could possibly know such a complex mathematical formula or how it could randomly select the duration of its fight. If the strategy prescribed by the mathematicians describes these animals' behavior correctly, then these seemingly absurd technical matters become meaningful. For the time being, our only question is whether this mathematical proposition correctly describes the behavior of the posing animals.

To answer this question, it is advisable to observe the real behavior of animals. The observed animals have no idea of our theory about them, nor that we are testing our theory on them. They simply fight to get what they want. We, however, can observe whether an animal poses against opponents of similar strength, under similar circumstances, for different amounts of time. Experiments have shown that the duration of posing in fact varies from fight to fight, and rather randomly. The next question is whether the duration of posing indeed conforms to the mathematical formula.

Such an investigation is more difficult. To make such a determination, we need to know the value of the given commodity for the given individual—and therein lies the difficulty. We cannot hope to get an exact value. However, we can give a rough estimation, on the basis of which we may at least see whether the animals' behavior is totally different from what the theory predicts. It is not. Even very rough estimates of value can predict the real behavior of the animals surprisingly well; not, perhaps, in any one situation, but in the long run.

Thus, when confronted with situations analogous to the dollar auction, animals behave more rationally than humans. They are unlikely to give more than one dollar for an object worth one dollar. Nor are they likely to give less. They usually pay just as much as it is worth. The weaker ones, the losers, should perish; they get nothing. It is the winners that sustain the species.

We humans have the possibility of obtaining essential goods more cheaply on the basis of sober judgment, conscious thinking, and perhaps mutual agreement. We have the capacity to agree without fighting, and where we cannot bargain, we can develop internal ethical principles that serve the good of the community better than brute force. Sometimes we really do this. At other times, however—as shown by the dollar auction—we find ourselves paying unrealistically high prices. It is as if the price of the ability to behave as ethical beings were the loss of our animal rationality.

2

The Brute as Hero

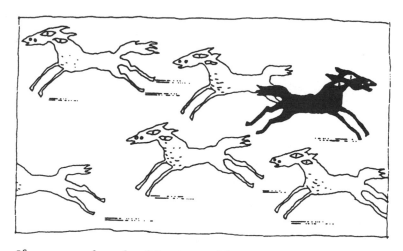

If everyone thought alike, it would mean the end of horseracing.

Douglas R. Hofstadter once proposed a competition to the readers of *Scientific American*, in which a prize of one million dollars was offered. Sort of. To put it more precisely, the amount of the prize depended in an inverse way on the number of competitors. If only one competitor entered, that individual would receive the full million dollars. But if there were two competitors, then one of them, determined by lot, would receive half a million dollars. If there were three competitors, then one of them, again by drawing lots, would receive

$333,333.33, and so on. If there were a million competitors, then the lucky winner would increase his or her net worth by just one measly dollar.

As in the dollar auction, this game—despite its simplicity—also has a special twist, making it amenable to modeling complex phenomena. It is a general characteristic of competitions for a set prize that the greater the number of competitors, the less chance there is for a single competitor to win, but the eventual winner gets the full prize, regardless of the number of competitors. Although every competitor decreases the others' *chances* of winning, the number of competitors does not lessen the *joy* of the winner.

In Hofstadter's game, however, each competitor reduces the joy of the winner, for had that competitor not competed, the winner would have received more money. This holds, paradoxically, even for the winner—by competing he reduces his own joy—although had he not entered, then of course somebody else would have won. In this game each competitor spoils the game. The only ethical solution seems to be that one should avoid such games by not competing. Every competitor—a.k.a. spoiler of the game—justly brings dishonor upon himself.

Yet if everyone thought this way and declined to compete in order not to reduce the great prize, then the chance of a big profit would be missed by everyone, who would then have only themselves to blame for everyone's missing out on the opportunity of a lifetime.

Games like Hofstadter's can be used to model how resources, natural and otherwise, can be utilized optimally, and how mankind squanders these resources. The role played by *Scientific American* corresponds to the resources present in the environment. Situations like the one modeled in the competition occur frequently in everyday life: There exist great, "million dollar" opportunities that vanish as soon as everyone tries to take advantage of them.

If there are only a few taxis in a city, then a few taxi drivers might become very rich in no time at all, but if everybody becomes a taxi driver, then no one will be able to make a decent living driving a cab. In such a case the city authorities may enter the fray and issue a limited number of taxi permits. In other cases, patent rights or copyright may ensure that the great chance remain really great, but only for the inventor of the idea.

Every year the Immigration Office of the United States draws lots for a few thousand green cards. The merits of the competitors are not considered at all; only blind chance determines who will win permanent residency in the U.S. Such a lottery resembles the solution offered by mathematicians to the problems of posing fights. The "land of unbounded opportunity" has come up with this solution in order to prevent the great chance from being diluted by masses of immigrants, to ensure that those who do immigrate will be able to make use of the great opportunity. Drawing lots is fair insofar as everyone competing for a green card has an equal chance of winning—there is no authority attempting to judge the real capabilities of the competitors—and anyhow, it is not at all clear what the competitors should be capable of.

If the competitors for a green card applied directly to the Immigration Office, their applications would probably be refused. Those applying in this way to immigrate would have to accept the authority of the agency and enter the competition. In the *Scientific American* game, however, there can be no competent authority. No legal articles determine who may or may not enter the competition. The only factor is individual decision, and as a consequence, those who enter the competition automatically spoil the game.

Dice and the Common Interest

A respectable person does not enter the competition announced by *Scientific American*. Yet the common interest of the readers demands that such a great opportunity not be missed. Otherwise, the *totality* of the readers would be a million dollars poorer than otherwise.

Common interest demands that somebody—despite the dishonor as a spoiler—enter the competition, but that only a single person do so. The question is whether such a *common interest* can exist, or whether it is just a dream. What kind of common interest is it if one person wins a large sum of money while the others look on politely and observe how their fortunate colleague has become wealthy? Why him and not me? Where is the *common* interest here?

If all the readers of *Scientific American* are truly governed by common interest (if such a thing is possible at all), then each person will decide whether or not to enter the competition based on the same line of reasoning. If everyone wants to serve the *same* common interest, then either everyone or no one will enter the competition. The result is the same: Nobody wins anything, and *Scientific American* laughs in its sleeve.

Thus, it is wise neither to enter nor not to enter the competition. But it is also impossible not to participate at all. Everyone who gives the least thought to the question of whether or not to compete has automatically become a player and has made some decision, even if only to exclaim, "This whole game is just a bunch of bull!" We have arrived at the same stalemate as in the dollar auction. Nor can one simply give up the game, because the fate of the goods at stake has to be decided one way or the other.

Once again mathematicians have come up with an artful suggestion to resolve the stalemate. The essence of the suggestion is that if there is no authority to draw lots, then each reader may do so himself.

Suppose one hundred thousand readers of Hofstadter's announcement consider entering the competition, thus automatically becoming players. Each player might decide whether or not to enter by employing the following procedure: Roll a die of one hundred thousand sides, and if it shows the number 100,000, you enter the competition; otherwise, you do not. Assume for a moment that every one of the hundred thousand readers follows this procedure. In this case we can make the following statements:

1. *Every player has an equal chance of entering the competition.* This chance is exactly one out of one hundred thousand for each player.
2. *Every player makes his decision on whether or not to participate on the basis of the same principle.* This way no one may reproach the winner with having behaved unethically.
3. *There is a good chance that there will be only one competitor, whose prize will be the greatest possible, namely one million dollars.*

Thus, the players as a totality make the best use of the great chance offered by Hofstadter's provocative offer.

Point 3 above is true, but the subsequent remark—as we will see—is not completely accurate. Nevertheless, it illustrates the basic idea of the solution: The common interest of the group of readers—that the prize of the winner should be the greatest possible—can be realized. Having found a *theoretical* solution through which common interest can succeed, it is no longer beyond reason to talk about common interest as a practical matter.

A Martian Humankind

Imagine for a moment what the world would be like if human beings in situations analogous to that posed by the *Scientific American* competition always made their decisions according to the above procedure. For the time being, let us forget issues such as the production of large numbers of fair dice of a hundred thousand sides. Let us assume that children learn it in kindergarten or that they inherit this ability genetically. (Even stranger mental games have led to the discovery of fundamental laws of nature. How absurd it must have seemed at one time that an iron ball might fall to the ground no faster than a feather. It is clear that the former falls swiftly, while the latter drifts lazily to the ground. Nevertheless, this seemingly absurd supposition laid the foundations of classical physics.)

How, then, could mankind become reasonable and equitable decision-makers as described above? Can we imagine a society in which a person who on throwing the dice and failing to roll a 100,000 actually refrains from entering the competition and does not, for example, roll the dice once more, saying, "The dice fell accidentally," or "The dice got stuck"? This appears to be an issue of personal ethics. Each person has to struggle with his or her own conscience. When it comes to breaches of social custom, society can sanction only strikingly serious infractions. Nature can help by developing a gene for conscience—if humans "equipped" with ethical principles and a conscience prove better at reproducing, then the rest will be taken care of by natural selection.

Were the principles of thought and behavior implied by the problem-solving method of casting dice to become general, this could have far-reaching consequences. Just imagine that somewhere, on Mars, for example, there is a society where the principle of casting dice has infiltrated Martian thinking deeply and irrevocably as an ethical imperative. There may be no *Scientific American* on Mars, but there are most likely environmental resources analogous to those represented by Hofstadter's offer. What would an earthling see who knows nothing about the ethical principles of the Martian psyche? It would appear to him as though there were many decent and honest individuals who refrain from entering the competition, while there was one brute—an aggressive individual—who has taken advantage of the integrity of the others to carry off the prize. Our earthling, ignorant of what was going on behind the scenes, would be greatly dismayed by the wickedness of this Martian competitor and would not understand that the other Martians would simply shrug it off.

These Martians are extremely effective. They are able to utilize the possibilities offered by the natural environment with the greatest efficiency, while simultaneously preserving it. Furthermore, our Martians probably are not particularly envious of the winner, since they know—they feel it in their bones, assuming they have any—that the fortunate winner did not amass a fortune through some skulduggerous manipulation. They feel deeply that the *common interest*—that someone, and preferably only one, utilize the great chance—was embodied *by blind chance* in that individual. It is the nature of the game that only one person can win maximally, and so it is completely in the common interest that the Martian who wins should win big. Today me, tomorrow thee.

However intelligent, effective, and fair this solution employing hundred-thousand-sided dice might be, it seems extremely unlikely that something even remotely similar could be realized in life as we know it. Human beings generally make their decisions by weighing the pros and cons and acting according to their emotions and moods. They very rarely allow blind chance to determine their actions. But perhaps this is the very reason why our moods, emotions, and susceptibilities are so variable. This may be how we can best approach the chance-based method of decision-making, which may well be the most rational. However, we have a long way to go before we see how that might happen.

The Protagonist Appears

The problem in taking advantage of the great opportunities that life affords, those like the one offered by *Scientific American* that are lost either if nobody jumps at the opportunity, or if everybody does, is not solved in nature by dice, but rather by *diversity*. Nature's dice are her fundamental genetic, quantum-physical, economic, and psychological mechanisms.

The role of the dice is filled by our moods, our occasional displays of courage, our uncertainties, and our spontaneous changes in perspective. Our Martian, who has real dice in his hands, does not in entering the competition have to take courage against the wrath of his compatriots, or against his own conscience in case he should cast the winning number: According to Martian ethical standards it is his duty to cast the die. For us earthlings, however, a variety of opposing forces fight it out inside us until we finally decide whether or not to compete.

It is not only that diversity exists among individuals of our species, but that there is a great diversity within every individual. To enter the competition goes against our ethical standards, and we may justly fear vilification. Nevertheless, the temptation is great, and who can know in advance what the result will be. The competitor may be despicable before the fact, but should he be the only one to enter the competition, then he will be hailed as a hero who has saved mankind from missing a great opportunity. Sometimes it is really blind chance that decides who will become a hero.

Game Theory

Game theory is a purely mathematical discipline that grew mainly out of the work of John von Neumann in the middle third of the twentieth century. The ideas of game theory are exemplified by the mathematically inspired solutions to the problems of the dollar auction and the million dollar game. But these are applications of game theory rather than examples of its theoretical basis.

John von Neumann believed in the power of reason. He thought that there must be a way of dealing with important situations in our lives in a purely rational way, at least with those situations that can be described in terms of abstract and unambiguous rules—like the dollar auction, the million dollar game, or such games as chess, Monopoly, and poker. Yet how firmly this belief in the power of pure reason was founded is not at all clear. It seems that to approach most such games perfectly rationally requires infinite lines of reasoning, on the order of, "I think that he thinks that I think that...." We feel in such cases that there can exist no perfectly rational method of playing such games. The best we can hope for is to carry such lines of thought as far as possible. It seems that a perfectly rational individual can never make a decision—that such an ability is vouchsafed only to those of limited rationality.

A less brilliant mathematician would probably have given up at this juncture, noted that there are many things in heaven and earth beyond our poor power to reason why, and returned to more traditional mathematical questions. Von Neumann, however, saw the possibility of creating a mathematical discipline out of this chaotic vicious circle, and he began to elaborate a new theory with complete mathematical rigor. He showed in 1928 that in quite a large category of games it is possible to play perfectly rationally, and that it can be done without infinite loops of thought, extrasensory powers, or perfect psychological intuition. All we need are some dice and a calculator.

We are not overstating the case. As we will see in Chapter 6, in many cases we really need nothing more than a pair of dice for playing perfectly rationally. The theory of games and its consequences have radically changed our views on the concept of rational thought, about the motives of human thinking, and, in fact, even about the causes and purposes of many kinds of diversity in the world.

Von Neumann's theorem founded a new branch of mathematics, one based on the possibilities of generalization and modeling of a variety of real-life situations. The theory of games that developed has proved to be a fertile and exciting mathematical discipline, and moreover, it has offered extremely effective methods for dealing with dilemmas of decision-making, conflict resolution, and social dilemmas. Let me just once invoke "authority": The 1994 Nobel Prize for economics was awarded to three outstanding scientists—J.F. Nash, J.C. Harsányi, and R. Selten—who work in this field. But the game-theoretic approach has proved to be fruitful not only in economics, but in biology, social psychology, political science, and a number of other areas as well.

As has happened in the wake of other great scientific discoveries, the terminology of game theory is increasingly infiltrating our everyday vocabularies, with terms like "non-zero-sum game," "mixed strategy," and "prisoner's dilemma" becoming part of our everyday language, as did "energy," "evolution," and the "unconscious" in an earlier era. My first sighting of a term from game theory in ordinary discourse was in the *Economist*, and I have seen more such in *Newsweek, Der Spiegel,* and the Hungarian *Heti Világgazdaság (World Economics Weekly)*.

Pure and Mixed Strategies

The fundamental concepts of game theory are usually explicated in textbooks written for students of mathematics, economics, and social biology, which, according to the usual format, contain many definitions and a rather profound mathematical apparatus. We are not going to follow that path, worthy as it is. Although we shall pay the price that the mathematical subtleties of game theory will remain obscure, we shall be compensated in that the basic ideas will probably become clearer, and we shall have a greater opportunity to explore applications of game theory to other branches of science, especially psychology.

As for the concept of "game" itself, we are going to use it totally intuitively. Thus, we are going to talk about two-person and multiperson games simply and naturally and about games of complete and incomplete information without defining them strictly. But we cannot avoid defining the notions of pure and mixed strategy.

A player is said to play a *pure strategy* if his actions are dictated by some *principle*, so that in identical situations the same action always follows from this principle. Someone who always, under all circumstances, unconditionally, observes the commandment "Thou shalt not kill!" is playing a pure strategy. A nonpure strategy is played by the soccer player who in similar situations kicks the ball now to one, now to another, of his teammates depending on his mood and intuition. In the *Scientific American* game, the thought, "I would be crazy not to enter the competition seeing that a million dollars is at stake," indicates a pure strategy. Persons who think like this will enter every such competition.

A *mixed strategy* is one in which the player first assigns a probability to each possible move and then decides on the basis of these probabilities how to proceed. *The actual decision is governed by chance*, but the probabilities for the different decisions are not necessarily equal. Players who decided to enter the million dollar game by tossing a die of a hundred thousand sides applied a mixed strategy with the following probabilities:

1. pure strategy: "I will enter" with probability 0.00001.
2. pure strategy: "I will not enter" with probability 0.99999.

The soccer player uses a mixed strategy if he directs a penalty kick to the left with probability 50%, to the right with probability 30%, and to the middle with probability 20%. More precisely, this player truly applies a mixed strategy only if he makes his decision by some random selection process. If this soccer player really believes in using a mixed strategy, then he should roll a ten-sided die. If it shows 5 or less, he should kick to the left; if it shows 6, 7, or 8, he should shoot to the right; and if it comes up 9 or 10, then he should aim down the middle. Only in such a way can he be sure of playing his chosen mixed strategy undisturbed by the goalkeeper's feints, his own mood, or the cheering from the stands.

Our solution to the dollar auction game—which turned out actually to have been adopted by certain animals—refers not to any moment of the game, but to the whole of it. The player does not decide from moment to moment whether to continue bidding but decides beforehand, using a mixed strategy, how far to go. Thus, mixed strategy may be applied both to a given move and to the whole game.

There exist strategies that are neither pure nor mixed. For example, when somebody decides, "I shall cast my horoscope once a year and plan my actions accordingly," he follows neither strategy. He does not play a pure strategy, because in the same situation the horoscope may dictate one decision this year and another the next. Nor does he follow a mixed strategy. If he did, then he might have to decide one way today and another tomorrow—yet according to his resolution he can make decisions only at the moment of casting his horoscope.

Game theory deals only with pure and mixed strategies. Taking guidance from one's horoscope is not unheard of and indeed is rather widespread. Psychology is interested in such modes of decision-making, since they are natural human functions. But since game theory was developed specifically to understand and investigate rational decisions, those based on horoscopes and similar methods are outside its sphere of interest. But mixed strategies do fall within the bounds of rationality, and the dollar auction and the million dollar game were our first examples of how mixed strategies appear in the theory of games.

Optimal Mixed Strategies

When we were speaking about mixed strategies, we nowhere said that it was forbidden to assign zero probability to certain pure strategies. That would have been silly. If a soccer player plays a mixed strategy, he is certain to assign zero probability to the possibility of kicking the ball into his own team's goal (whether he can do this is another question altogether). Thus, a player using a mixed strategy may assign zero probability to all but one move and 100% to the move that remains. This will then be a pure strategy, and thus pure strategy can be considered a special case of mixed strategy, or equivalently, mixed strategy is a generalization of pure strategy.

In the million dollar game, if all one hundred thousand players use a mixed strategy and all roll a hundred-thousand-sided die, then it could happen that two or more players throw a 100,000, so that the winner gets only half a million dollars or even less. But if nobody rolls 100,000, then *Scientific American* will get away without paying a cent—which may easily happen. The probability of nobody throwing a 100,000 is about 37%!

If the aim of the players is to realize their common interest, then all of them want to maximize the winner's gain. For this, the optimal strategy is not to cast a hundred-thousand-sided die. If the number of sides is somewhat less, then the probability of two or more players entering the competition will be greater, but in compensation the danger of nobody entering the game will be less. If we continually decrease the number of sides of the die while calculating the expected loss to *Scientific American*, we find that this loss will increase for a while, but then a point is reached where the probability of more than one player entering becomes so great that the journal's expected loss begins to decrease. It can be calculated that the expected loss to *Scientific American* is greatest if the players use a die of 64,532 sides. The optimal mixed strategy of our imagined Martians is to roll such a die, and anybody who throws a 64,532 will enter the competition.

(The actual winning number could be anything—but each player has to determine his own winning number in advance.)

This strategy can be considered an *optimal* mixed strategy, because the expected gain of the totality of the players will be greatest this way. Since only one person can win a given game, in the long run, *every* player optimizes his winnings if this strategy is adopted by all.

How can our Martians know how many sides their dice should have in a given game? Even if the concept of mixed strategy, and the consequent ethical principles, were deeply embedded in their psyches, we cannot expect each of them instantly to calculate the optimal size of dice for a given situation.

There are two answers to this question. First, it is enough if only one well-educated specialist calculates the number of sides and informs the others of the result. The transmission of information must be trustworthy, of course, and the other Martians must believe it implicitly. The second answer is given by the theory of evolution. Of the competing species of Martians, those that will survive are those best able to select the optimal size of dice. They can ensure this by educating a few specialists and developing trustworthy mass media or by developing a sufficiently high level of general mathematical intuition in every member of the species. This mathematical intuition may be based on simple rules like the one that can be drawn from the million dollar game: "If there are many of us and if preferably only one of us should be selected, then we should use a die whose number of sides is about two-thirds our number."

Humanity did something like this when it dragooned almost all of its members into many years of mathematical studies in school. Still, it was probably more important for us to develop other kinds of intuition. Our mathematical intuition is rather fallible. Even if we now believe that 64,532-sided dice give a better overall expectation than 100,000-sided dice, there are very few of us who intuitively felt this and expected such a solution at the outset.

Who Optimizes and Why?

If we look only at the good of the community, then in the million dollar game the optimal mixed strategy may not be the only one that leads to the best result. In the end, it is enough if there is always precisely one player who chooses to enter the competition. This can be achieved by other means than individual mixed strategies, even without the introduction of an authority. Imagine a gene that has developed two alleles, a competitive one that compels an individual to enter a competition, and a noncompetitive one that forbids him to enter. If the resources for survival can be obtained only through million dollar games, then a population with no competitive genes will sooner or later die out, because its members never access vital resources. But a population with too many such genes will also be unsuccessful.

The population most likely to survive will be one that preserves some of its competitive genes but that develops in its nonplayers a capacity for exacting more or less successfully a ransom from the winners. The method and degree of ransom could be determined by other genes—but this lies outside the scope of our investigation. In order that there can be something of value to be shared, the resources provided by nature must first be utilized successfully.

We can also talk about the optimal mixed strategy if the competitive behavior is actually determined by genes. In this case, however, it is not the individual that plays the optimal mixed strategy but nature itself, when it develops the two alleles and distributes them at random throughout a population. Only the mechanism of evolution can develop a species in which the frequencies of the two genes approach the proportion of the optimal mixed strategy.

Obviously, our model is too abstract. It is not very realistic to suppose that the resources of survival are distributed only in situations resembling the million dollar game. There are many games through which one can obtain resources for survival, and each of them may have a different optimal strategy. Most of these games, however, share the feature that a mixed strategy is more effective than any pure strategy.

In short, a mixed strategy can be played by the individual or by nature, both of which can mix their strategies in many ways. The individual may roll a die, using mathematical intuition to select one with the optimum number of sides, or he can spontaneously change his moods, attitudes, and priorities. Nature can develop a diversity of individuals through different genetic patterns, or it can produce beings in whom different strategies of behavior constantly compete.

3

The Prisoner's Dilemma

If you can see bars in front of you, it does not mean that you are a prisoner. You may be on the outside looking in.

The prisoner's dilemma is the "rubber bone" of game theory—it can be chewed over forever. Thousands of mathematicians, psychologists, political scientists, philosophers, and economists have thought about it, trying to find a solution. Yet it remains as mysterious and baffling as in 1950, when Merrill Flood and Melvin Drescher first proposed it. The name *prisoner's dilemma* was given by Albert W. Tucker, who in 1951 wrote the first paper about it. Tucker presented the problem in the form of a short detective story. Here is one version:

The police apprehend two criminals who together have committed a serious crime. There is no direct evidence that these two men actually committed the crime, and all the police can actually prove is an instance of excessive speeding. The prosecutor would very much like to close the case, and therefore she makes the following proposal separately to the two prisoners, whom she has placed in separate cells:

"Here's the plea bargain: If you will confess to the crime and implicate your accomplice, thereby helping us to resolve this case, I will set you free, and we'll forget about that little matter of speeding. In this case your accomplice will be shut up in prison for ten years, and the whole matter will be closed forever. This offer is valid, however, only if your accomplice does not confess to the crime and thus does not help us in clearing up the matter. If he also confesses, then, of course, your confession is not of much value, as we will know everything without it. In this case, each of you will be jailed for five years. If neither of you confesses, we shall, alas, be unable to convict you, but we'll be very severe on that rather nasty incident of speeding, and you will both be imprisoned for one year. Finally, I must inform you that I made this very same offer to your accomplice. I await your answer tomorrow at ten o'clock. Just think—you can be free by eleven!"

Let's summarize the situation in a table:

| | | the other accomplice | |
		confesses	does not confess
the first accomplice	confesses	**-5**, -5	**0**, -10
	does not confess	**-10**, 0	**-1**, -1

The first number in each cell, the one in boldface, indicates what the first accomplice gains, while the second number shows the gain of the other accomplice. Since it is worse to be jailed for ten years than for five, years of imprisonment must be considered a negative gain, whence the minus signs. In the given situation the best achievable result is zero, being set free.

Two Logical Solutions

The two accomplices, we shall assume, have no commitment to each other; their alliance was completely accidental. Each of them has the sole aim of getting off as cheaply as possible. What is the most logical action: To confess or not to confess?

Imagine yourself in the shoes of one of the accomplices. If my accomplice confesses, here are the alternatives: If I also confess, then I will be jailed for five years. But if I do not confess, then I'll be imprisoned for ten years. Thus, if my accomplice makes a confession, then it would be better for me to confess, too. Now, if my accomplice does not confess, there are again two alternatives: If I confess, then I'll be free tomorrow. But if I do not confess, then I'll be in prison for a year. Thus, if my accomplice does not make a confession, then it would be best for me to confess.

No matter what my accomplice does, I'm better off if I confess. There is no other choice. Logic demands a confession.

Logic dictates the same thing to the other prisoner. Thus, as both of them are logical beings, they will both confess—each receiving a sentence of five years, although they could have gotten away with only one year apiece if neither of them had confessed. This is the *prisoner's dilemma*. The question is whether this logic is somehow mistaken. In other words, does logic exclude the rational cooperation of the two prisoners?

The following line of reasoning is just as logical as the previous one: I do not like my accomplice particularly, and I have no attachment to him. Nevertheless, swine though he is, I know that he is just as intelligent and logical a person as I am. Otherwise, I would not have made an alliance with him. I also know that he is in an identical situation. He is not attached to me either, and furthermore, he received the same offer as I did. Like me, he will base his decision on his own interests and on logic.

Logic is always guaranteed to lead to the same conclusion. Two and two are four, no matter by what method a correct calculation is made. Thus, my accomplice will make the same decision as I will, no matter what it will be. Consequently, if I decide to confess, then I am sure that he will arrive at the same conclusion. If my line of reasoning leads to not confessing, his thinking will lead to the same result.

If I decide to confess (and thus so will he), then I will get a five-year sentence (and so will he, though I do not care about that part of the business—tough luck for him). If I decide not to confess (and so will he), then I will be in jail for one year. It is better to be imprisoned for one year than for five. So I will not confess.

This line of reasoning seems to be just as logical as the previous one. But how can two equally logical arguments lead to opposite results? Is one of them false? Or is there a problem with logic?

On the Nature of Logic

Careful comparison of the two lines of reasoning will reveal that the second one includes a type of step that the first one does not. It assumes that two complementary lines of reasoning necessarily lead to the same conclusion. Only this step could cause the contradiction.

Here is the solution. What really follows from the two lines of reasoning *together* is that prisoner's dilemma situations cannot exist at all. To be more exact, this follows solely from the initial statement in the second argument. The first argument applied only steps that are also allowed in the second argument. Therefore, we can continue the second argument by simply adding the first one to it. This way we arrive at a contradiction. Consequently, according to the rules of logic, the hypothesis that there is a prisoner's dilemma must be impossible.

If prisoner's dilemmas do not exist, then anything can be deduced from them if they are assumed to exist, since it is a rule of logic that anything follows from a false statement. If witches do not exist, than the statements "All witches ride on broomsticks" and "No witches ride on broomsticks" are logically equally true. This may sound strange, but it is merely vacuously true to assert that every one of the zero witches that exist rides on a broomstick. On the other hand, if we do not construct logic this way—so as to accept such vacuous statements as true—then logic itself will be inconsistent and therefore useless. That's just how logic works.

If prisoner's dilemmas do not exist, then in a prisoner's dilemma situation it is logical both to confess and not to confess. Our first argument happened to arrive at the former conclusion, the second at the latter.

As Gödel's theorem (1931) showed, there is no system of any complexity in which all of the truths that can be phrased within the system can be proved within that system. The first line of reasoning merely used the usual rules of logic. This does not mean that the additional condition of the second argument—that two logical arguments must lead to the same conclusion—is false. Nor does it exclude its falsity. Neither can be deduced within this system.

Logic in its usual form does not include this additional condition. Thus, this system really excludes the cooperation of the two prisoners within the prisoner's dilemma but does not exclude the existence of a world in which no prisoner's dilemmas exist. For the time being, this is only an abstract possibility, but in Chapter 4 we shall see concrete examples of the *mechanisms* by which a world free of prisoner's dilemmas can be realized. In such a world the prosecutor could not put the two prisoners in such a situation. Even if she could, the two prisoners simply could not perceive the situation as a prisoner's dilemma. Let us consider an example.

Merrill Flood, one of the discoverers of the prisoner's dilemma, once proposed the following deal to a secretary at his institute: He would give her $100, or, alternatively, he would give her $150 if she and her colleague could agree on how to split the larger amount between them. Flood offered no money to the other secretary. Flood was interested in learning in what way and on what basis the secretaries would divide the extra $50. To his great surprise, the two secretaries soon came forward for $75 each, even though the first secretary could have easily gotten $100 without even mentioning the offer to the other secretary. The secretaries, however, saw the situation as a cooperative game, completely avoiding the competition for the extra money.

For those who think this way, prisoner's dilemmas really do not exist, even if there is a prosecutor who would like to create one. Although logic prohibits that the prisoners cooperate if prisoner's dilemmas were to exist, it does not exclude a world in which there *can be* no prisoner's dilemmas. Perhaps it would be nice to live in such a world. We know from experience, however, that prisoner's dilemmas do exist in our world.

Everyday Prisoner's Dilemmas

Two gas stations are located next to each other. At the beginning of each month each owner must set the price of gasoline for the coming month, state law prohibiting changing prices during the month. The new price has to be posted exactly at midnight on the first day of the month.

The owner of one of the stations might think thus: I made a small profit at last month's price, but not much. If the other station were to go out of business, my sales would double, and I would then make a huge profit, since the costs of maintaining my station would rise very little. It would be worth a little sacrifice to take customers away from my competitor. What if I lowered my price a little? Although I would make a little less profit on each gallon of gas, my sales would almost double, and this would be well worth it.

After complicated calculations, the owner of the gas station ascertains that if he lowered his price a certain amount and thereby won over half of his competitor's customers, then his one unit of profit would rise to four units. But then doubt begins to gnaw at him: What if the owner of the other station also thought this way and also lowered his price? Then my sales wouldn't go up a bit! Worried, he makes new calculations and determines that next month his business would bring him zero profit with the lowered price. Thus, it is not worth lowering the price. Now that doubt has awakened within him, he feels compelled to perform additional calculations: What would happen if he kept his old, higher, price and the other owner lowered his price? The outcome would be disastrous! Maintenance costs are so high that with only half the sales volume, his deficit would be three units even at the higher price.

Midnight is approaching, and if he wants to set a new price, he must post it as the clock strikes twelve. Just in case, he has prepared a new board with the lower price. If he sees his neighbor lowering his price, he can quickly do the same. Slowly he walks out to his price board as midnight begins to strike, and he can see that the other owner is walking to his price board, also looking worried, clutching a small board. The two are just about to enter into conversation when they notice the dreaded Inspector of Municipal Order, who is watching what will happen to the prices at midnight. There is no time for negotiations, which anyhow are illegal. Both owners have to decide immediately whether or not to change the price or leave the old one. The critical moment is at hand—the stroke of midnight— and neither can see what the other is doing.

The situation can again be summarized in a table:

		second owner	
		lowers the price	does not lower the price
first owner	lowers the price	**0**, 0	**4**, –3
	does not lower the price	**–3**, 4	**1**, 1

The logic is exactly that of the prisoner's dilemma: No matter what the second owner does, the first is better off if he lowers the price. If the second owner lowers the price, then the first owner can avoid a loss. If the second owner does not lower the price, then the first owner can quadruple his profit. Thus at midnight, greed and fear of loss demand a lowering of the price, but if they both do it, they will both lose all their profit.

Simple buying and selling can also lead to a prisoner's dilemma, especially on the black market, where there is no guarantee that we will find each other the next day. There isn't much time for checking: I can buy with counterfeit money; the seller can give me defective goods. Once we have our purchase, no matter what it is, we are better off if we paid with counterfeit money. Once the seller has the money, he is better off if he gave us shoddy goods. But if we both do this, nobody wins anything, although both of us could gain something in an honest transaction.

Puccini's opera *Tosca* demonstrates a typical prisoner's dilemma. Tosca's lover, Cavaradossi, has been sentenced to death by the corrupt captain Scarpia. Scarpia, however, is greatly taken with Tosca, and offers her a deal: If she sleeps with him, Scarpia will order the firing squad to fire blanks. Tosca tells Scarpia that he can have her only after he has issued an irrevocable order to the firing squad. Tosca, however, does not choose the cooperative solution: During their embraces she stabs Scarpia to death. It soon becomes apparent that Scarpia, too, has not chosen the cooperative solution. The order was counterfeit, the volley is fired, and Cavaradossi is killed. What did we expect from opera anyhow? The logic of the prisoner's dilemma prevails.

The logic of the arms race also suggests a typical prisoner's dilemma. Balance could develop between the two opposing superpowers either if both sides armed themselves to the teeth or if both of them armed themselves only moderately. The cheaper balance of power is clearly better for both parties than an expensive equilibrium. Now our table looks like this:

		the strategy of the other superpower	
		arm to the teeth	arm moderately
the strategy of the first superpower	arm to the teeth	**2**, 2 (expensive balance of power)	**4**, 1 (superiority)
	arm moderately	**1**, 4 (defenselessness)	**3**, 3 (cheap balance of power)

The numbers show the order of desirability: A score of 1 represents the worst possible outcome in the situation, while 4 is the best. An expensive balance is better than defenselessness, and superiority is better than a cheap balance. This order of values can be questioned, and it should be, but many believe in it, especially if superiority can be turned into direct economic advantage. Game theory assumes that the players are clearly aware of their own (perceived) interests and values. It is not the task of game theory to change them, but by its very abstractness, game theory can call attention to the necessity of change—for example, when it clearly points out that a certain system of values inevitably leads to a prisoner's dilemma.

The prisoner's dilemma is principally about cooperation, its evident necessity and frequent near-impossibility. In all of our examples, one of the strategies involved cooperation, while the other did not. The prisoner who does not confess, the gas station owner who does not lower his price, the superpower that does not arm itself excessively, are cooperative. With this behavior, if both parties think similarly, a better result can be achieved. The noncooperative strategy will be called *competitive*, although this word does not always express the essence of the thing. It is not quite suitable to apply this term to Tosca.

Prisoner's Dilemmas with Many Persons

The above examples have shown that cooperation usually means giving up something. Thus one can easily find oneself in a prisoner's dilemma. The formula is the following: Take a temptation that, if everybody succumbs to it, leads to catastrophe. But that is not all. A special arrangement of values is also necessary. There are other severe dilemmas to which the conclusions drawn from the prisoner's dilemma cannot be easily applied. The million dollar game does not function this way, although there was a temptation there, too, and if everybody had succumbed, no one would have profited. The difference lies in the fact that in the prisoner's dilemma, the competitive player causes harm to cooperative players, while in the million dollar game the competitive player causes harm only to the other competitors.

The prisoner's dilemma with many persons is also called the *problem of common pastures*, which is modeled by the following example: A village has a common pasture. There are ten farmers who have cows, one each, and all ten cows graze on the common pasture. There they wax fat, thereby more or less eating up the meadow. The farmers get richer, and one or two of them can soon afford a second cow. When the first farmer sends his second cow to pasture, little change is noticed. Perhaps slightly less grass can be eaten by any one cow; perhaps one fatted calf will become less so. When the second and third farmers send *their* second cows to pasture, still no great problems arise. Although the cows become visibly thinner, each remains well-fed and healthy. But by the time the seventh farmer gets around to buying a second cow, all the animals are hungry, and the total value of the seventeen cows is less than that of the original ten. By the time the ten farmers have two cows each, all twenty cows have starved to death. Throughout this process, two cows are always worth more than one, and so it is always advantageous for a farmer to buy a second cow—until they all starve to death.

From this description it is apparent that we have found ourselves once again in the logic of the prisoner's dilemma. But beware, not all social dilemmas are prisoner's dilemmas. Let us have a look at the common pastures "game" table:

	the majority	
	buys another cow	does not buy another cow
buy another cow	**2**, 2 I have two very thin cows	**4**, 1 I have two rather fat cows
do not buy another cow	**1**, 4 I have a very thin cow	**3**, 3 I have one very fat cow

myself (to the left of the bottom two rows)

Once again, the numbers represent the degree of benefit: The best case scores 4, the worst scores 1. The second numerals in the table indicate how the others fare in the given situation *on average*. A precise analysis of the game would require a more complex table that shows whether each of the ten farmers is cooperating or competing. That complex table is summarized in this small one, in which the behavior of only one farmer is emphasized. The numbers in the table are precisely those in the arms race table, and therefore the logic here is that of the prisoner's dilemma. This table is valid until all the cows have starved to death. When this happens, the numbers in the table change, but by then it is too late to profit from the knowledge that the prisoner's dilemma has been operating again.

A panic is a typical example of a prisoner's dilemma with many persons; when fire breaks out in a crowded room, for example. There is a special case, one that has often occurred, when the door of the room opens inward. The cooperative behavior would be for everyone to step backward a couple of steps, allowing the door to be opened easily, and then everybody would be saved. This generally does not occur: Everyone runs to the door, crushing one another to death.

Iterated Prisoner's Dilemmas

The original story of the prisoner's dilemma describes a particularly critical situation in that I have only one choice, after which everything will be over. If I, as one of the accomplices, do not cooperate and my partner makes the mistake of trying to cooperate, then he will not have to reproach me for at least ten years. If he, too, does not cooperate, then he'll have nothing to reproach me for when we get out of jail in five years.

The situation is slightly different when we face the same partner repeatedly. In this case, if we do not cooperate at the outset, we sentence ourselves to eternal competition, because the partner we cheated once will probably never trust us again.

The case of the two service station proprietors is an *iterated prisoner's dilemma*, since on the first day of next month they will meet the same dilemma, unless the unilaterally cooperating partner goes bankrupt in the meantime. When watering lawns is prohibited during a drought, it is another case of an iterated prisoner's dilemma. The cooperative behavior is to obey the order, while it is competitive behavior when somebody secretly waters his lawn, achieving personal gain but endangering the whole community's water supply. There are similar examples to be found in relation to pollution of the environment.

The logic leading to competition is not complete in a situation with repetition. If there are many rounds, the choice is not only between cooperation and competition. Complex long-term strategies are available as well. For example, one possible strategy is that at first I cooperate, but if my partner does not reciprocate, then I will never cooperate again. Another possible strategy is to cooperate all the time, hoping thereby to make my opponent see reason sooner or later. Or I can cooperate every other time, regardless of my partner's move. The possibilities are infinite.

Prisoner's dilemmas do not arise only in human interactions. Sticklebacks exhibit very interesting behavior when a large fish approaches. They cannot tell in advance whether the large fish wants to make a meal of them. It would be simple to flee from every large fish, but then stickleback life would be one of constant flight, and they would have no time for other vital activities. On the other hand, the fatalistic solution, "Let's see what this big fish will do," can also be dangerous, since an unexpected attack could destroy a whole school of sticklebacks. Consequently, the sticklebacks do the following. A reconnaissance group gradually approaches the large fish. They swim toward it, stop for a while, approach, swim a few more inches, stop, approach, and so on. If they get so close that the large fish could easily attack them and nothing untoward happens, they return to the other sticklebacks and continue their usual activity. If, however, the large fish catches one of the reconnaissance group, then the rest rush back to the others to sound the alarm.

The prisoner's dilemma arises within the reconnaissance group. One or two sticklebacks may chicken out and turn tail. Those who return will individually be safe, but if they all flee, then all of the sticklebacks may fall victim to the large fish, including the deserting scouts and their offspring. If the remaining reconnoiterers do not flee, they will be more endangered individually than before the few fled, because each one's chances of being caught will have increased if the large fish turns out to be a stickleback eater. The logic of the situation matches exactly that of the problem of common pastures. We will soon see how sticklebacks cope with this dilemma.

Axelrod's Competitions

The American political scientist Robert Axelrod has investigated the problem of whether cooperation can theoretically develop in a world where everybody is governed by his own interests. In 1979, Axelrod asked several well-known scientists, many of whom had published papers on the prisoner's dilemma, to participate in a competition. He asked them to send him the strategy they considered optimal for the iterated prisoner's dilemma. Axelrod asked for the strategy in the form of a computer program. He then had the programs play against one another in a round-robin competition: Each program played a series of two hundred iterations against each of the others. In each of the two hundred rounds, the programs received scores according to the following table:

		program 2 cooperates	program 2 competes
program 1	cooperates	**3**, 3	**0**, 5
	competes	**5**, 0	**1**, 1

The winner of the round-robin competition would be the program with the highest cumulative score. Axelrod did not reveal the number of rounds in advance so that this information could not be programmed into the competing strategies.

Fourteen programs entered the competition, from the very simple to the highly elaborated. The most complex of these attempted to outsmart their opponents using concepts borrowed from the field of artificial intelligence. To these programs Axelrod added a fifteenth, which cooperated or competed at random. The winning program, that of the famous social psychologist Anatol Rapoport, was the simplest of all. It consisted of just two rules:

1. Cooperate in the first round.
2.   Thereafter, do whatever the opponent did in the previous round.

Rapoport named his program "Tit for Tat." It is abbreviated TFT in the game-theory literature.

What is so ingenious in this embarrassingly simple strategy that it was able to defeat the highly wrought programs of the best experts?

The Personality Traits of the Programs

Axelrod was able to study psychological concepts in a way that had never been possible before. Unlike human beings, each of these programs could be analyzed to see how it worked, and thus the degree to which each program conformed to certain psychological concepts could be studied, provided that the psychological concept was sufficiently well defined. Thus the relative survival values of the various "personality traits" could be assessed in terms of how they performed in the competition. What personality traits are useful in an iterated prisoner's dilemma?

Axelrod found two such traits that were invariably associated with the highest-scoring programs. The first was *niceness*. Axelrod defined this concept as never initiating competition. It does not mean that a nice program never competes, but that it never begins a competition. The other concept was *forgiveness*. Axelrod considered a program forgiving if after a lapse into competition by its opponent it was willing to return to cooperation, provided that the opponent did so as well. In other words, the program didn't hold a grudge. Almost all of the programs in the leading half of the field included these traits, but none of the others did. The winner, TFT, featured both characteristics.

Axelrod next tried to discover a strategy for beating Tit for Tat. He found three, one of which was neither nice nor forgiving. But another superior strategy was one that works exactly like TFT except that it does not retaliate immediately, but only after two successive competing moves. These results cast doubt in two directions: On the one had, it was now unclear whether niceness and forgiveness are the traits best suited for survival in an iterated prisoner's dilemma, while on the other hand, there appeared to be an advantage in being even more "forgiving" than TFT.

To help decide these questions, Axelrod announced a second competition. All of the competitors knew the results of the first competition and Axelrod's analyses. It promised to be an exciting tourney: Each competitor knew that niceness and forgiveness were good, but the very logic of the prisoner's dilemma suggested that an unfriendly and unforgiving program might easily take advantage of an overly "nice" environment. But then again, everyone knew this.

Sixty-two programs from six countries representing at least eight fields of science arrived for the second competition. Axelrod also entered the three programs that were capable of beating TFT.

Anatol Rapoport sent his TFT program again, and he won again! The program that was twice as friendly as TFT finished twenty-first; the unfriendly program that would have won the first competition ended up in the second half of the field.

Rapoport's social-psychological intuition worked remarkably well. There was certainly no guarantee that TFT was going to be a winner in the second competition, and the intuition of other entrants did not suggest this either. The success of TFT greatly depended on the other strategies. For example, TFT never wins a one-round competition because it always lets itself be exploited first, and it always forgives the opponent's competitiveness as soon as the opponent returns to cooperation.

Axelrod again examined the personality traits of all the programs. *Niceness* and *forgiveness* excelled again: They appeared in fourteen of the top fifteen programs, although they showed up in fewer than half of the programs altogether. Axelrod found three additional traits that were positively correlated with success. One was *provocability:* If the opponent competes, the program is very likely to retaliate. Another was *responsiveness:* The program responds in some way to its opponent. The last of the new successful strategies was *transparency:* Axelrod simply measured this by the length of the program—the shorter the program, the greater its transparency. Although computer science knows more subtle and intricate measures of determining the complexity of a program, this simple measure served Axelrod's purpose.

TFT carried all five of the traits to the greatest possible extent. Still, the almost completely pure demonstration of these traits was an important finding, because despite its simplicity, TFT carries many other traits that do not contribute particularly to success in an iterated prisoner's dilemma (although they are not harmful, either). The importance of these five traits was proven not by TFT, but by all the other programs. TFT's superiority probably lies in its ability to condense these five traits to a remarkable degree. While this may seem self-evident, Axelrod found that other cooperative personality traits were missing from many of the leading programs but were present in the programs that weren't quite up to snuff. Such traits proved not to be carriers of success.

We should honor Anatol Rapoport not only for his double victory, but for having the courage to listen to his intuition once again and resubmit the same annoyingly simple strategy. Yet Rapoport warns us not to overvalue TFT. He thinks that sometimes TFT reacts too violently to insults; there are many social situations in which Tit for Tat proves to be too harsh. For instance, after some "innocent" misconduct, an opponent may find itself hopelessly mired in competition. This impasse may be avoided only if the players occasionally include an "out of turn" pardon in their strategies.

Theoretically, it is easy to improve TFT. It may be sufficient to write a program that usually plays TFT but also continuously examines whether the opponent is reacting to its moves. If the opponent does not react (for example, it cooperates and competes at random), then TFT should switch to a constantly competitive mode, since this is the most effective strategy against such an opponent. Quite a few such programs participated in the competition, but they did not fare particularly well. Such programs are neither nice (they are willing to initiate competition against a noncompeting opponent) nor transparent (they appear to be TFT for a long time). Although with our present knowledge it is not surprising that they were not successful, it is nevertheless strange that such highly refined intelligence is so ineffective. The five simple traits discovered by Axelrod seem to be more effective than pure reason in a world full of prisoner's dilemmas.

The main lesson to be learned from Axelrod's results is that it is not impossible for cooperation to develop in a purely selfish environment. Once the gene for TFT (or at least the above five personality traits, including provocability!) develops in an organism, then this organism will be able to develop stable cooperation even if its aims are purely selfish.

Tit for Tat in Sticklebacks

The reconnaissance group of our stickleback acquaintances gradually approaches the large fish. The reason for this slow approach may be to make the prisoner's dilemma continue for several rounds rather than only one, because it is more difficult to develop cooperation in a single round, a situation in which Axelrod's results are not valid. Observation shows that there is indeed cooperation in a reconnaissance group, and that competition—that is, retreat—appears infrequently. It is unknown how sticklebacks have evolved cooperation. Do they play TFT? Or do they perhaps employ some other strategy?

The German ethologist Manfred Milinski designed an ingenious experiment to answer this question. He placed a stickleback in a large, rectangular aquarium. While the stickleback was swimming at one end of the aquarium, Milinski put another aquarium with a large fish in it at the other end of the stickleback's aquarium. He simulated a companion for the stickleback by placing a mirror along the long side of its aquarium. The stickleback, presumably having no idea that its apparent companion was a mirror image, headed toward the large fish. Thus, the first move was cooperation, as prescribed by TFT. Not surprisingly, the "companion" followed the stickleback in tandem. So far, the experiment models the situation of cooperation. The mirror, however, was movable, and the experimenter sometimes rotated the mirror forty-five degrees, with the result that when the stickleback approached the large fish, its companion appeared to swim away—it was not cooperating. The stickleback then retreated as well. When Milinski set the mirror so that the companion appeared to move first toward, then away from, the large fish, the stickleback followed the TFT strategy rather accurately, although there were exceptions. Sometimes, despite the "treason" of the companion, the stickleback cautiously approached the large fish alone, as sticklebacks sometimes do when they are alone and face potential danger.

Psychological Experiments with the Prisoner's Dilemma

It is astonishing how many important psychological concepts can be modeled by a simple formula like the prisoner's dilemma. The conflict between common and individual interest, trust and treason, greed and fear, revenge and forgiveness—all can be interpreted within this framework in pure form. No wonder social psychologists took to the prisoner's dilemma as geneticists did to *Drosophila* (the genetic makeup of the fruit fly can be mapped quite accurately. Its characteristics are simple enough to be isolated but complex enough to allow general conclusions to be drawn).

Well over a thousand papers on the prisoner's dilemma have been published. The results are contradictory. Experimental conditions varied greatly from laboratory to laboratory. Experimental subjects played for tokens, honor, or even large sums of money. The inconsistency of the results may be due to such differences in experimental conditions. Nevertheless, some general conclusions may be drawn.

The researchers freely varied the values in the table, thereby controlling the degree of temptation to compete (the gain of the person who competes against a cooperating person) and the price of being a "sucker" (one who tries to cooperate against a competing person). The results generally conformed to common sense. When the value of temptation was raised, the number of cooperative responses decreased, and the ratio of cooperative responses similarly decreased when the loss of the "suckers" increased. Thus, the inclination to choose the competitive strategy in a prisoner's dilemma seems to be determined in equal measure by greed and mistrust.

The researchers also varied the possibility of communication between the partners. Every possibility of communication increased somewhat the chance of cooperation. If the partners could converse about the game and their options, or if they were even allowed to come to an agreement, the ratio of cooperative responses increased, but only moderately. (Such agreement, of course, was not at all obligatory. Everybody had to decide alone whether to cooperate or to compete.) Communication made players aware not only of their common interest, but also of feelings of temptation and defenselessness.

The investigators also studied the effect of the subjects' personalities. These results were the most contradictory. Some researchers found that certain personality traits (especially traits similar to those found by Axelrod) tended to increase the probability of cooperation; others found no such relationships. Sometimes, when the game involved groups with a previous history, the subjects played out the dominance relationships within the group. Dominant persons in the group tended to be more competitive, and submissive members more cooperative.

Men were more likely to cooperate in this game than women. In a classic iterated prisoner's dilemma, almost 60% of the male pairs cooperated, while fewer than 35% of female pairs cooperated. Mixed pairs cooperated about half the time, but here there was no difference in the number of cooperating men and women; that is, there were as many men cooperating and women competing as the other way around. This result is particularly interesting if we compare it with the dollar auction, where women entered the bidding spiral less often than men. These findings are still a subject of debate. Some researchers have found that such gender differences depended on the gender of the experimenter. With women experimenters gender differences decreased. Gender differences increased if the loss of the "sucker" was increased. This may indicate that in the case of the prisoner's dilemma, women consider competition a cautious strategy for avoiding the worst outcome.

In the one-round prisoner's dilemma the overall proportion of cooperation was about 40%. It is a matter of taste whether this proportion is seen as high or low. One interpretation is that cooperation is fortunately not a rare bird, even if logic dictates otherwise. The contrary interpretation is that pitifully few people act in the common interest.

You might think that in an iterated prisoner's dilemma, cooperation would become significantly more frequent. In Axelrod's colorful words, the future casts its shadow before itself. The possibilities are more subtle; there is the option of revenge, but also of benevolence. In games of several rounds cooperation did increase, but it never rose above 60%. The game frequently ended with both players mired in competition.

In games of several rounds, the TFT strategy often appeared, but never in its pure form. Perhaps this is good, because the advantages of TFT appear only if it is played perfectly from the beginning. For instance, if the players happen to decide to play a pure TFT after mutual competition, they will get stuck in competition, and only an out-of-turn act of forgiveness can help.

Although sticklebacks do not play TFT in its purest form, they have reached a very high level of cooperation. Sticklebacks are capable of much more rational behavior than *Homo sapiens*, not only in the dollar auction, but in the prisoner's dilemma, too.

The Significance of Phrasing

The chief aim of social psychologists in conducting their prisoner's dilemma experiments was to learn how people can be most effectively moved to cooperate. Among the innumerable variations of the experimental conditions, an artful phrasing of the situation proved the most fruitful.

The prisoner's dilemma can be introduced as follows: "By pressing this button, that is, if you cooperate, you give 2 units to your partner and give 1 unit to yourself. By pressing this other button, you give 2 units to yourself and none to your partner. Your partner has the same choices." This is summarized in Table 1. Table 2 shows the original prisoner's dilemma, with the numbers indicating the order of advantage of possibilities: A score of 4 is the best result, 1 the worst. This table is the same as the one above for the arms race.

Table 1		
	to the	
	to oneself	partner
cooperation	1	2
competition	2	0

Table 2		
	the other	
	cooperates	competes
the one cooperates	**3**, 3	**1**, 4
one competes	**4**, 1	**2**, 2

If the numbers in Table 1 are considered as concrete scores, then our present game is exactly the prisoner's dilemma. For instance, when "the one" player cooperates and "the other" competes, then "the one" gives himself 1 point and gives his partner 2 points, while "the other" player gives himself 2 points and gives "the one" nothing. Thus "the one" gets 1 point and "the other" gets 4 points, just as in Table 2. The situation is the same in the other cases, too.

This rephrasing of the prisoner's dilemma makes players see *the exact same game* completely differently. We can also say that the two games are *logically isomorphic*; that is, logically we cannot differentiate between the two, since if a logical line of reasoning leads to cooperation (or competition) in one game, it will necessarily lead to the same behavior in the other. The fact that logically the two games are indistinguishable does not mean that they are psychologically the same. One of the games may elicit significantly more cooperative responses than the other.

Table 3 shows a variant of the same game:

Table 3

	to oneself	to the partner
cooperation	0	3
competition	1	1

You can easily calculate that this game is another version of *exactly the same* prisoner's dilemma. It is also logically isomorphic to the prisoner's dilemma.

Experiments have shown that this form of the game elicits significantly more cooperation from the players than the original prisoner's dilemma, while the one shown in Table 1 elicits less. Interpretation of the results is again a matter of taste. Perhaps the last table is so effective in promoting cooperation because it demonstrates very clearly that we can win something only if the other gives it to us, that is, if the other cooperates. This version of the game makes it most difficult for the players to avoid cooperation.

4

The Golden Rule

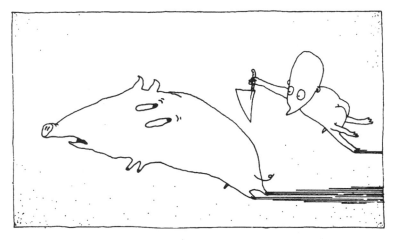

If you are doing this and I am doing that, we may be doing the same thing.

It is not a particularly heartening world we live in. We have been unable to find a reassuring solution even to such a simple problem as the prisoner's dilemma. The rephrasing of the problem at the end of the previous chapter may be witty, and even sometimes successful, but it does not take the sting out of the prisoner's dilemma. The possibility of treason and competition is ever present and must be reckoned with at every step.

Instead of helping us to find an acceptable solution to the prisoner's dilemma, logic has led to a bad solution, to the antithesis of cooperation. The only glimmer of hope was the recognition that logic does not exclude the possibility of a world in which prisoner's dilemmas cannot exist. This may be a logical possibility, but it is still unclear what real mechanisms could make the world free of prisoner's dilemmas.

The most obvious mechanism would be if the thought that universal rationality necessarily led to cooperation operated with compelling force in everyone. Such reasoning, however, cannot be perfectly universal, because no rule within a system of logic can govern how that system is to be applied. Such rules would make logic itself unable to function.

Perhaps giving up logic itself would not be too high a price if the resulting world, besides excluding prisoner's dilemmas, kept other useful human achievements. However, such a world without logic would be problematic, for logic lies at the foundations of mathematics, science, and even most of philosophy.

Fortunately, milder principles still compatible with logic make it possible to eliminate prisoner's dilemmas and all their painful consequences. The oldest such principle is the golden rule of the Chinese philosopher Confucius, who flourished around 500 B.C. Versions of this rule also appear in the works of Plato, Aristotle, and Seneca. The Gospel puts it this way: "Whatsoever ye would that men should do to you, do ye even so to them" (Matt. 7:12).

If this rule pervaded our thoughts so deeply that nobody could think otherwise, then everybody who met a prisoner's dilemma would think, "I would like to be treated so that I could be set free, immediately if possible, because that would be good for me. Therefore, I shall choose the solution that grants freedom to others. Therefore, I shall cooperate, and I will not confess." But if we want to preserve logic as well, then the contradiction persists, since the line of reasoning leading to competition remains valid.

Logic and the Golden Rule

Our ability to solve the prisoner's dilemma in contradictory ways means *either* that the reasoning leading to the golden rule is not logical *or* that no prisoner's dilemma can exist. If the reasoning based on the golden rule is logical, then *logically* the impossibility of prisoner's dilemmas follows, but this is only a formal game. We would like to understand intuitively why the golden rule excludes the prisoner's dilemma.

Prisoner's dilemmas work their wicked ways because we assume that people's values conform to their own interests. This priority is turned on its head by the golden rule, which can also be put this way: Consider the good of others as your goal. If *both* players follow this dictum, then the results of the prisoner's dilemma will be as in Table 1 below (as a reminder, Table 2 shows the values of the original prisoner's dilemma):

	Table 1 Golden Rule Values				**Table 2** Original Values	
	the other does not confess confesses				*the other* cooperates competes	
does not confess	**3**, 3	**4**, 1	*the*	cooperates	**3**, 3	**1**, 4
confesses	**1**, 4	**2**, 2	*one*	competes	**4**, 1	**2**, 2

My interest now dictates that the other become free immediately (I score 4), while second best is if he is sentenced to only one year (I score 3), and so on. Logic now dictates that I not confess, since no matter what my partner does, I am better off. The table no longer represents a prisoner's dilemma, since different values are assigned to the individual strategies.

Thus, if the golden rule is added to the rules of logic, then the game ceases to be a prisoner's dilemma: Individual interests and the common interest coincide. In this special case, logic works. But we also know from logic that the golden rule terminates the prisoner's dilemma not only in this case, but in *every* case: If the golden rule exists, then no prisoner's dilemmas exist, because such situations are no longer prisoner's dilemmas. For a person whose logic includes the golden rule, not even the wickedest heart could create a prisoner's dilemma. Merrill Flood's secretaries offer an example of how such a humankind would think. Furthermore, such beings, free of prisoner's dilemmas, would be just as rational as we are (if indeed we are rational at all), since all the other rules of logic remain unchanged.

Moreover, a humanity that built its house upon the rock of the golden rule would not have to give up any of the achievements of mathematics, science, or philosophy.

But first we must clarify a slight inaccuracy. The golden rule is not a rule of logic. It is not a *form of logical reasoning*, but rather it determines the *logic of value selection*, something that does not usually fall under the purview of traditional logic. Traditional logic, that is, the study of correct forms of reasoning, is here being supplemented by an additional, ethical, principle. This principle does not contradict traditional logic, but neither does it follow from it. Logic applies with equal force to a world where the golden rule prevails as to one in which it does not. In the latter world, logic concludes that cooperating is not worthwhile. In the former, logic concludes that prisoner's dilemmas do not exist.

The Categorical Imperative

In our world, the logic of the golden rule does not prevail unconditionally, and this may not be due only to our fallibility. As George Bernard Shaw said, "Do not do unto others as you would that they should do unto you. Their tastes may not be the same." Even if the golden rule excludes the existence of situations like the prisoner's dilemma, it does not exclude the existence of other dilemmas. In fact, as we shall see later, there are dilemmas in which the application of the golden rule is definitely disadvantageous to all participants.

Immanuel Kant suggests an even more general solution in his *Foundations of the Metaphysics of Morals*. Kant takes as the basic concept of his ethics a fundamental law of practical reason that he calls the law of the categorical imperative. He stated this law in many forms, of which perhaps the most often cited is the following: "Act as though the maxims according to which you live should by your will become universal natural law." A maxim is a strictly personal rule of conduct that determines in *every situation* what the individual should do. Kant calls such a rule *categorical* if it prescribes the *same* behavior in every possible situation. The Ten Commandments are maxims (for example, Thou shalt not kill), as are all significant ethical rules. According to Kant, a maxim can serve as a principle for general laws if its application does not lead to logical inconsistency.

Kant then states that "The pure concept of the categorical imperative also gives the formula into our hands." It is very rare in the world of ethical questions that the strictly formal analysis of duty leads to precisely definable behavior. This is what makes the concept of a categorical imperative so powerful.

In Kant's philosophy, the law of the categorical imperative stands above all other laws. He wrote: "There are two things that amaze me; the starlit sky above me and the categorical imperative within me." The maxim of the categorical imperative "begins to speak" in a fully rational being without a guilty conscience. According to Kant, it has always been present in human nature, in all ages and in all societies.

Kant introduces many illuminating examples of the operation of the categorical imperative and the consequent ethical laws. His examples, however, are not particularly convincing for today's readers. Kant uses the concept of logical inconsistency without the strictness that prevails today. Let us try to apply Kant's method to the prisoner's dilemma.

Let's see. Can I possibly want competition to be the universal maxim? If so, then I should be prepared to be sentenced to five years in prison. On the other hand, if I want cooperation to be the universal maxim, then I will be sentenced to only one year in prison. Thus, competition cannot be prescribed by the categorical imperative. Cooperation does not lead to a similar inconsistency, and therefore it is the behavior to be adopted.

This argument strongly resembles the one at the beginning of the previous chapter that led to our choosing cooperation. In fact, they are the same argument. We used the concept of logical inconsistency very loosely, as Kant did. As a matter of fact, we cannot prove *in general* by the methods of logic that an assumption does not lead to logical inconsistency. This is a mathematical consequence of Gödel's theorem. The categorical imperative cannot be applied as a general logical formula. Still, we may find it useful. Perhaps this is why we can easily understand the reasoning leading to cooperation, even though it is not strictly logical. Our understanding consists of more than logic.

Although we cannot prove in general that a maxim does not lead to inconsistency, in special cases, when the number of choices is finite, we may be able to do so. This is the case with the prisoner's dilemma. The question arises again, Why does it follow from the categorical imperative and from logic that prisoner's dilemmas do not exist?

The answer is different from the one the golden rule led to. Unlike the golden rule, the categorical imperative does not change the order of values. By conceiving some decision as an *ethical* issue, we excluded from the beginning any solution whose result was asymmetrical, under the assumption that the categorical imperative functions in everyone and that the question is perceived as an ethical one by all. In the world of ethical issues, that is, in a world where the categorical imperative can really operate as an ultimate, unconditional law, a prosecutor simply cannot set up a dilemma for the two prisoners such that "If you do this and the other does that, then you'll be jailed for this long and he'll be jailed for that long." This would be just as meaningless as if he proposed to set them free if each is both a man and a woman.

The Categorical Imperative and Diversity

What do the golden rule and the categorical imperative have to say about the million dollar game?

The prescription of the golden rule is clear-cut: It would be good for me if I won a million dollars. Therefore, I must act so as to make it possible for others to win. Therefore, I will not enter the competition. Since I certainly cannot injure anybody in this way, I have nothing further to think about. Naturally, if everybody abides thus by the golden rule, nobody will enter the competition, and the great opportunity will be lost forever. But the golden rule must be worth this trifling loss. It is not necessary to grab every opportunity.

According to the categorical imperative, neither entering the competition nor refraining from entering is correct, since both pure strategies (we might say both possible maxims) lead to nobody winning anything in a situation where somebody could win something. The alternatives lead to "logical" inconsistency. Kant's original form of the categorical imperative cannot suggest an ethical solution to this game.

We, however, already know other possibilities. We have seen that with the application of a mixed strategy the game can be solved justly and with equal chances for everyone. Can a mixed strategy be considered ethical?

The application of a mixed strategy is incompatible with the golden rule, since according to this rule the only ethical behavior is not to enter the competition. However, the application of a mixed strategy is not necessarily incompatible with the idea of the categorical imperative. If I *want* a mixed strategy to be a universal maxim, I do not have to arrive at a logical inconsistency.

Taking it further, if I wanted everybody to play according to a mixed strategy that is different from the optimal mixed strategy, then the expected gain would be less, which is logically inconsistent with the condition that I want a general principle by which the winner could gain the maximum. The optimal mixed strategy, however, is not inconsistent with any other possible mixed or pure strategy (this is why we call it optimal). Thus, I can safely desire that everybody act according to this maxim.

This deduction is logically incomplete, because as we saw earlier, there are strategies that are neither pure nor mixed. However, we have also seen from the beginning that the categorical imperative is incompatible with the complete system of logic. Thus, we can ignore this little spot of logic. Now we can say that because the optimal mixed strategy conforms to the law of the categorical imperative, it can definitely be considered ethical.

This reasoning has strange consequences. The optimal mixed strategy sometimes leads to entering the competition, sometimes to refraining from entering. Thus, both entering the competition and refraining from entering are ethical. Something must be wrong here.

The question of entering or refraining from entering is a mere superficiality. The essential matter is that we honestly roll the dice and act according to its dictates. Thus, those who entered the competition and those who did not acted according to the same maxim, provided that the categorical imperative was working in all of them. In Chapter 2 we talked about imaginary Martians in whose psyches it was permanently and firmly engraved that they make such decisions according to the optimal mixed strategy. The *mechanism* of such resolve could well be the categorical imperative. These Martians could resemble us earthlings after all, for the categorical imperative as an ultimate ethical principle may be active in them.

Our idea of "categorical" has changed considerably. Kant considered a maxim categorical if it prescribed the same behavior in every possible situation. This definition remains valid, but meanwhile the nature of behavior has changed. Behavior is no longer what somebody visibly does. Behavior now means that we throw the dice and act accordingly, though this cannot be seen from the outside. What is visible is only the consequences of rolling the dice. Thus, the principle of the categorical imperative is not necessarily incompatible with acting differently in identical situations. Our behavior may be governed in both cases by the same strict ethical principle.

A biological analogy (of which Kant could have had no more idea than of game theory) may illuminate this point. Biologists speak of *genotype* and *phenotype*. *Genotype* refers to the genes carried by the organism, and these genes largely determine the organism's external characteristics. *Phenotype* refers to the sum of the specific external characteristics, whose possibilities are greatly limited by the genes but are not completely determined by them. Phenotypes develop as an interaction between the genotype and the environment. In our analogy, the maxim corresponds to the genotype, which cannot be seen. Behavior is the phenotype, which can be observed. The categorical imperative in this analogy is the interaction between genes and environment that determines the phenotype.

The analogy is not as haphazard as it might seem. Game theory has found its deepest, furthest-reaching applications in biology. One of the reasons is that the biological interests of an organism can be reasonably expressed by a single number. It wants to maximize its own survival value. In this number, such qualitatively different values as food and freedom are combined into a single quantity. Another reason is that in biology, the mechanisms of inheritance have been studied in great detail, giving us a rather exact picture of the mechanisms—such as DNA—that can realize the abstract games of game theory. We know much more about genetic machinery than about the functioning of ethical judgment.

The Gift of the Magi

It is morning, and a young couple are engaged in a heated argument about their evening plans. The young man proposes a basketball game, while the woman would rather go to a concert. They have no time to reach an agreement, and they run off to work without reconciliation. They cannot talk to each other during the day. They both leave work shortly before seven o'clock in the evening, at which point each has to decide—individually, without communicating with each other—where to go: to the basketball game or to the concert.

To make this a problem of game theory—let's call it the "gift of the Magi" after the O. Henry story—both players must have clear-cut preferences. As a first choice, both the man and the woman want to spend the evening together; the desire to spend it at the preferred activity is only a second-place choice. For both of them, the worst outcome is to spend the evening separately, especially if the woman spends it at the basketball game and the man at the concert. This alternative is worth only 1 point for each of them. It is somewhat better if they spend the evening separately, but at their preferred venues (2 points to each). For the woman, it is better if they are together at the concert (4 points), and only slightly worse if they attend the basketball game (3 points). It is the same for the man, *mutatis mutandis*. Thus, the table of the game is as follows:

| | | *the man* | |
		to the basketball game	to the concert
the woman	to the basketball game	**3**, 4	**1**, 1
	to the concert	**2**, 2	**4**, 3

The table reveals that this is again a dilemma. Let us see what the ethical laws have to say.

According to the golden rule, the woman thinks, "I will go to the basketball game, because it is good for the other. The man thinks, "I will go to the concert, because it is good for the other." Thus, the worst possible outcome occurs. Yet were they to follow the golden rule exactly, each must think, "I would like to be treated in such a way that my partner goes where I want to go. Therefore, that is where I shall go." By this logic, the man goes to the basketball game and the woman to the concert. They still spend the evening separately, but at least more pleasantly.

Thus the two different interpretations of the golden rule, "Do what is good for the other" and "Do what you would like others to do to you," do not necessarily lead to the same result. Yet the usual assumption that the two interpretations are equivalent is not incorrect; we need only look at the individual strategies at a more abstract level. We have to look at them from the heights of ethics. Each person has a selfish and an unselfish alternative. The selfish alternative for the man is to attend the basketball game; for the woman it is going to the concert. Now, applying the golden rule literally, both of them may think, "I would like the other to use the unselfish strategy with me. Therefore, I will use that, too." Thus the golden rule, the law of goodness and altruism, leads to the worst possible outcome. Shaw's "inverted golden rule" is in effect here.

Perhaps it is too frivolous to treat this trifling disagreement as a grave ethical problem. Nevertheless, let's have a look at the effect of the categorical imperative in Kant's original interpretation. As we have seen, we can consider the asymmetric cases as nonexistent, leaving us with two cases. Between the two, it is better for "me" (either for the man or for the woman) if I choose the selfish strategy. Accordingly, the categorical imperative requires that the man go to the basketball game and the woman to the concert. This is better than the result dictated by the golden rule, but it is far from optimal.

What does game theory have to say about this? If we think in terms of a mixed strategy, then the man and the woman will each cast a die and make their decisions accordingly. The only question is the number of sides of the die. In other words, what should be the probability of going to the preferred place? It can be calculated that the two of them together will achieve the greatest expected score if the man goes to the basketball game with a probability of 5/8 and to the concert with a probability of 3/8, and if the woman goes to the concert with a probability of 5/8 and to the basketball game with a probability of 3/8.

Taking into consideration all the possible outcomes, their expectation is to gain a total of 5.125 points. Although this is considerably less than the 7 points they would obtain if they spent the evening together at either place, it is much more than the meager 2 points they would get by application of the golden rule or the 4 points achieved under the sway of Kant's original categorical imperative.

Like the solution of the million dollar game, this solution is also compatible with the principle of the categorical imperative, provided that both parties roll the appropriate die and comply with its dictates.

The Basic Forms of Two-Person Games with Mixed Motivation

John von Neumann's revolutionary conceptual abstraction was the realization that any game whose system of rules can be unequivocally described and in which the players are fully aware of their interests can be expressed in number tables. Such tables can be refined: Perhaps the two service station proprietors play the prisoner's dilemma game on the basis of different profits in summer than in winter, or perhaps the expenses of the two gas stations are different, so that the game is not completely symmetric. Perhaps the man would feel less out of place at the concert than would the woman at the basketball game. Von Neumann discovered the conceptual and mathematical means to treat such values; the concept of mixed strategy also comes from him.

Given such an effective conceptual apparatus, it is worth examining the basic forms of our everyday games and social dilemmas. What tables express various strategic situations, and what methods can be found to treat the principal forms of games? Life, of course, does not create such pure situations as the prisoner's dilemma, but once we know the basic forms of dilemmas and conflicts and their mechanisms, we won't be groping so much in the dark, noticing the trap only after we have fallen into it.

Sticking to two-player games, and even imposing the restriction that the players have only two possible moves, we ask first what basic forms we can find. We already know two types: the prisoner's dilemma and the gift of the Magi. Are there other situations with fundamentally different mechanisms of operation?

To answer this question, we shall continue to study the *order* of "pleasantness" of the possible outcomes, disregarding specific *degrees* of pleasantness. This means that we have to review all tables where the scores 1, 2, 3, 4 appear in different combinations for the different players. There exist 78 essentially different such tables, 12 of which conform to the condition that the players be in a symmetric situation, but only 4 of which can be considered dilemma situations. Here is a nondilemma game:

		moves by player 2	
strategy		A	B
	A	4, 4	2, 3
moves by player 1			
	B	3, 2	1, 1

Evidently, both players should choose strategy A. Otherwise, both would be worse off. This way they automatically reach the optimum, without conflict.

We already know two forms of the four possible symmetric two-person "dilemma" games with two moves. Here is an overview of the tables of all four: *coop* stands for "cooperative," *comp* for "competitive."

	Prisoner's Dilemma		Gift of the Magi		Leader		Chicken	
	II.		II.		II.		II.	
	coop	comp	coop	comp	coop	comp	coop	comp
coop	3, 3	1, 4	1, 1	3, 4	2, 2	3, 4	3, 3	2, 4
comp	4, 1	2, 2	4, 3	2, 2	4, 3	1, 1	4, 2	1, 1

(each grid labeled "I." on the left side)

The game of *leader* closely resembles the gift of the Magi, but here mutual cooperation does not lead to the worst outcome: Mutual competition is even worse. The game resembles what transpires when two excessively polite persons motion each other through a door. Here, the competitive strategy is to insist that the other go first, since each player wants to earn points for politeness. The cooperating person, willing to accept being viewed as impolite by the other, is prepared to go first. The worst solution occurs if they both compete, because then both will starve to death in front of the door. It is somewhat better if both cooperate and collide in the doorway, because at least they squeeze themselves through at the expense of a little discomfort. If one of the players competes and the other cooperates, they both get through easily, but the competing person is slightly better off, because in addition to getting through rather quickly, he can also despise the other for his impoliteness. The story may be a little artificial, but the game itself is not particularly interesting from a psychological or game-theoretical aspect beyond ascertaining that such things do in fact happen. The game of *chicken* is much more interesting.

Chicken

Chicken got its name from the American movie *Rebel without a Cause*. In this film, produced in 1955 (with many inferior successors), Los Angeles teenagers played the following game: They would drive in cars toward a cliff. Whoever jumped out last was the winner. A variant is the so-called chicken race. Two boys driving stolen cars speed toward each other on a narrow road. The first to swerve out of the way is the chicken, despised by the whole gang.

If we look at the advantages of outcomes in this game, we arrive at the table for *chicken* above: For me, the best result is to hold out to the end (I compete) and for the other to veer out of the way (he co-operates). It is somewhat worse for me if we both get out of the way, because although I remain alive, the two of us are chickens. But it is better to be a chicken than to have a head-on collision.

Actually, the duel with cars is in effect a game of many rounds, since both players must decide at every moment whether to get out of the way. There comes, however, a moment of truth. If neither you nor your opponent gets out of the way, there will be a collision. At this crucial instant the players have to make their decisions independently of each other, without knowing what the other has decided—the game thus proceeds in accordance with the table of *chicken*.

This is the essence of the chicken games encountered in life: Although the final decision is a single one, it is preceded by a longer or shorter "prelude," and the decisions of the players depend heavily on conclusions drawn from this prelude. If one of the players convinces the other that he will not give way under any circumstances, then the other will be forced to step aside to avoid the worst. The means of persuasion are many. As Hermann Kahn writes in *On Escalation* (p. 11), "The 'skillful' player may get into the car quite drunk, throwing whiskey bottles out the window to make it clear to everybody just how drunk he is. He wears very dark glasses so that it is obvious that he cannot see much, if anything. As soon as the car reaches high speed, he takes the steering wheel and throws it out the window. If his opponent is not watching, he has a problem; likewise if both players try this strategy."

This strategy may not be the most rational, but it probably serves the purpose. If both players adopt it, then the outcome is surely fatal, but there is nonetheless a risk in not playing this strategy, because that permits the opponent to play it. The more irrationally one plays the game, the more likely it is that he will win. The situation was different in the prisoner's dilemma: No matter what the opponent did there, I was better off if I competed. Rationality dictated competition, even if I knew what the other would do. Here, however, if my opponent competes, it is advisable that I cooperate, but if my opponent cooperates, I am better off if I compete. If I do not know what the opponent will do, rationality does not give any definite advice. Irrationality, however, can help us to persuade the opponent that the only rational solution for him is cooperation.

In a chicken game, the only chance for mutual cooperation is if each party makes it clear to the other that cooperation on his part is out of the question. The player who cannot bear the risk of the worst outcome is a certain loser in such games. And these games are very common.

Before World War II, England's prime minister, Neville Chamberlain, was unwilling to risk the worst—war—and at the outset Hitler won quite a few chicken-type games against him. It was Churchill who recognized the situation as a chicken game and forced England into the war.

During the Cuban Missile Crisis in 1962 President Kennedy's advisors analyzed the situation by game-theoretical methods and demonstrated the chicken nature of the conflict early on. This enabled Kennedy to make it clear to the Russians that he was not willing to compromise on this issue. He succeeded in persuading Khrushchev that the United States would not shrink even from nuclear war. In the end, it was Khrushchev who turned the steering wheel and veered aside.

Game theory is an abstract discipline dealing with rational decisions. The power of the theory can be seen in the fact that paradoxically, it was able to prove that sometimes the only possible rational behavior is irrationality.

It was fairly easy for us to review all the possible combinations of two-person games with two choices and to ascertain that there are four dilemma situations among them. If we now give each player three choices, then there are almost two billion different games. Nobody so far has felt the urge to write them all down, especially since such an effort seems unlikely to uncover any radically new ideas. The basic dilemma mechanisms have been demonstrated by the above four games. Real-life conflicts usually consist of complex combinations of these four forms.

Asymmetric Games

Asymmetric games arise when the players are in different situations, or if their situations are symmetric but their priorities differ. Perhaps the difference is small: Maybe the man is slightly sorrier to miss the basketball game than is the woman to miss the concert. In this case the game is not very different from its symmetric counterpart. The basic mechanisms are only slightly altered by differences in evaluation. But sometimes the interests of the players differ radically, and in these cases even the preferential ordering of the possible outcomes may differ. For example, one of the players may perceive the game as chicken, the other as a prisoner's dilemma.

King Solomon once had to do justice in the following situation: Two women each claimed to be the mother of a child. One of them was the real mother, the other an impostor. King Solomon raised his sword, proclaiming that he would solve the problem by cutting the child in two, with half an infant going to each woman. One of the women immediately renounced her claim. King Solomon declared the child hers.

If we have the wisdom of Solomon, perhaps we do not need game theory. But we can arrive at his solution by game theory even if we happen to be endowed with a lower wisdom quotient. The values of the real mother are different from those of the impostor: The game is asymmetrical. For the mother, the game works according to the logic of chicken; furthermore, she sees that the opponent is determined to compete. For the impostor, the worst outcome is the other woman getting the child; it is somewhat better if the child is killed. Her values are those of the prisoner's dilemma. Each woman's behavior is predicted by game theory. King Solomon's wisdom lay in recognizing the original asymmetry of the game and finding a situation that led to predictable differences in behavior.

The Dollar Auction and Ethical Principles

The dollar auction resembles both the prisoner's dilemma and chicken in many respects. At each turn, a player can decide whether to stop bidding, which is a cooperative solution, or to continue, which means competition. However, it can be debated which is worse for me: if I do not bid and the other gets the money, or if we both bid and the price of the dollar climbs to vertiginous heights. In the short run, the former may seem worse, because it cannot be foreseen how high the bid will go. In this case, the dollar auction operates as a prisoner's dilemma. But if we know the mechanism of escalation, it turns out to be more like chicken. The behaviors observed in real life are closer to chicken.

Horrified as we may be at the less than fifty percent cooperation observed in the prisoner's dilemma, the dollar auction is much worse: The level of cooperation is minuscule. Even if the game stops at fifty cents—which, as we have seen, indicates uncommon self-restraint—quite a few turns of competitive bidding have likely preceded this single instance of cooperation.

The tit-for-tat strategy has proved to be a rather stable solution to the prisoner's dilemma for both mankind and sticklebacks. In the dollar auction, however, this strategy leads to catastrophe if even a single bid is made. My opponent will outbid me, and I in turn will outbid him—there is no surcease in tit for tat.

You might say that even if only one person makes a bid, it is no longer tit for tat, since such a strategy is not "nice." But as opposed to the prisoner's dilemma, if nobody bids first, the chance for a 99-cent profit is missed. Perhaps the person who bid one cent should not be considered competitive, since he only warded off the danger of everybody missing a great opportunity. But the other bidder, who would have "sacrificed" himself just as gladly by giving one cent for a dollar, does not see it this way—and he is right. As children often say, "It all started when he hit me back." Such beginnings almost always lead to games of dollar auction type.

That is how the nuclear arms race began. The individual steps of the game can be conceived as a prisoner's dilemma, but the nuclear arms race *as a process* worked as a dollar auction. If one side gained an advantage over the other through some technical invention or by spying, the other party bid higher, until one of the parties collapsed. To make matters worse, the initial step—the USA developing the atomic bomb—should not be condemned in this context, since it was deployed not against the Soviet Union, but against their common enemy.

The golden rule tells us definitively not to bid in dollar auctions, since you want others not to bid against you. The golden rule prevents us from falling into such a trap, but it also prevents the community from reaping a profit, just as in the million dollar game. In the dollar auction it would be theoretically possible for the players to draw lots among themselves to determine who should bid that single cent. However, this cannot be done in reality, because the rules of the game—as in any auction—forbid collusion among the players. The auctioneer would immediately intervene.

The categorical imperative if applied only to pure strategies would echo the golden rule: Do not bid, because if bidding were the general maxim, everybody would lose by it, and this is logically inconsistent with the aim of avoiding loss.

But if we take mixed strategies into consideration too, then the categorical imperative suggests something different. If there are two players, then the following is better than a maxim against bidding. In every instance of bidding toss a coin: Heads you bid, tails you don't. If both players act accordingly, they are highly unlikely to have to pay more than a few cents for the dollar, because a long string of heads is improbable. It remains to be answered, of course, why the players should bid with a probability of 50%. Isn't the probability too great that nobody will bid, causing the great chance to be missed? The question of what strategy will optimize the players' expected gain is not, however, an ethical issue, but a mathematical one.

We have simplified the game a little in that we did not differentiate between the first bid and subsequent ones, although at the outset both players are free to bid, while in subsequent rounds only one player decides whether to continue. The following maxim might seem better than the previous one: "Use a mixed strategy to determine the first bid, but do not make the second bid under any circumstances." The players would thus give no more than one cent for the dollar, and so maximize their joint profit. In this case, however, matters outside the sphere of ethics will settle such issues as who gets the money if both players happen to toss heads. The mathematical model of the game can be refined, but the result will always be the same: If we look for a universal maxim that is valid for everyone and is based only on ethical principles, then it can lead to the bidding of 2, 3, or even more cents with a certain (rather low) probability.

Cooperation and Rationality

In the chicken game, as in the prisoner's dilemma, both the golden rule and Kant's categorical imperative lead to mutual cooperation. However, neither of these ethical laws leads to the best solutions in the gift of the Magi and leader. In fact, these two ethical principles may lead to opposite results. The radical difference between them can be seen in that mixed strategies can be easily adapted to the categorical imperative but not to the golden rule. If two people have the same the system of values, then the golden rule excludes the possibility of two equally ethical persons acting differently, while the categorical imperative does not.

Ethics has long been a difficult concept from both philosophical and practical viewpoints. We may consider a maxim ethical only if it leads to cooperative behavior. Game theory, however, has shown that the concepts of rationality and cooperation are rather unclear.

Look, for example, at our earlier table of the four basic games. In the prisoner's dilemma, in leader, and in chicken, cooperation meant that a better result could be achieved if the two players acted similarly. In the gift of the Magi, however, this is not the case. You might, of course, simply exchange the labels of the two strategies. But then "cooperative strategy" would mean that one follows one's own selfish interests. This sounds strange. No matter how we define cooperation, we could probably find a game in which the given concept of cooperation leads to an absurd result.

Rationality is even worse. Pure reason *cannot possibly exist.* A game, after all, is only a table of numbers, and we can find games conforming to any kind of table. *We can create a table of numbers (that is, a game) for any concrete concept of rationality such that the given rationality concept leads to complete failure for both players.* It is a logical consequence of Gödel's *mathematical* theorem that such a table of numbers can really be constructed, even if we allow both players to use mixed strategies. You tell me a concept of rationality, and I tell you what game we'll play. Knowing Gödel's theorem, I will certainly be able to find a game where if we all play according to your concept of rationality, we'll all lose, although if we played according to another concept of rationality we'd all win.

Prisoner's dilemma is a concrete game that questions the traditional concept of rationality based on pure reason. We could take the sting out of this in two ways: When we included either the golden rule or the categorical imperative in our system of thought, the prisoner's dilemma ceased to exist. But then the gift of the Magi caused a mess; there, the golden rule was of no help, but the categorical imperative—with the aid of mixed strategies—was. But we already know that this cannot be the ultimate form of rationality. To solve new problems, we shall always be forced to find new forms of rationality.

5

The Bluff

Some yogis can pierce a long needle through their chest. But it's just a trick: First they push their hearts out of the way.

Bluffing is a special form of lying or deception. Bluffers make statements, show behaviors, and perform activities that would be perfectly all right if they were not completely unfounded. A poker player will call if he has good cards, but he may also call if he is bluffing. Bluffs can be deceptive because the opponent may draw erroneous conclusions from them—that the causes that would normally produce such behavior are really there.

In bluffing, no untrue statement is actually made. The bluffer says nothing specific about the object of deception, for instance, about his actual cards. In fact, the essence of bluffing is the inexpressive behavior itself, the so-called poker face. After doing what reality does not justify, the bluffer acts as if he had just done the most natural thing in the world. In everyday life, bluffs are not always so clear-cut. Sometimes bluffing involves showing off, talking big, or a lot of eyewash. Here we will use the word in its pure sense.

An unprepared student is bluffing if he looks at the teacher as if he knew the whole lesson, but it is a simple lie if he says openly that he did his homework. The outcomes of the two strategies differ when the teacher finds out that the student knows nothing. In the case of the bluff, if the next time the student looks as intelligent as previously, this will provoke the teacher to challenge him—enabling the student to dazzle the teacher and profoundly change her opinion of him (unless, of course, he was bluffing again). But an open lie, if detected, will not provoke us to make another test. Once the teacher labels a student a liar, she will not ask that student again whether he has prepared the lesson. As the boy who cried wolf teaches us, a liar is never believed, even when he is telling the truth.

Whole libraries could be filled with the literature on lies, but bluffing is very rarely mentioned. This is all the more interesting since bluffs work completely differently from other forms of lies, and most statements that are true of lies prove incorrect where bluffs are concerned. Lies, for instance, should be ethically condemned under the categorical imperative, as Kant remarked. Lies violate the categorical imperative because if everybody lied continually, this would logically contradict the notion that statements have meaning and that conclusions can be drawn from them.

Kant's idea can perhaps be debated, since lies generally contain information from which true conclusions might be drawn. But this could be done far more reliably without lies.

Bluffs, on the other hand, applied in an appropriate proportion, can lead to an equilibrium that is optimal for everybody, and nobody's interest would dictate changing it. There are situations in which we can find a maxim (based on mixed strategy) that satisfies the categorical imperative only through occasional bluffing. Certain types of *optimal* mixed strategies necessarily contain bluffs, and while similar situations may arise with other types of lies, it is with bluff that this reasoning can be studied in its pure form.

Bluffers have different motivations from those of liars. Generally, the aim of a liar is to make others believe his lie and to have the environment rearranged accordingly. The liar profits from this directly. A real bluffer does not mind if his bluff is called, because the next time, when he is not bluffing, he can gamble for really high stakes. Someone who never bluffs will never get this opportunity, because as soon as he makes a large bet, his opponents will immediately assume that he has a strong hand and will not risk much.

A bluff can sometimes win, of course, but for a born bluffer this is only a secondary gain: It only decreases the cost of bluffing. The chief gain in a bluff that has been revealed lies in leaving doubt about future bluffs. He who bluffs for immediate gain is no different from a liar, and in the long run he will suffer the same fate: Once he is unmasked, everything is over. Real bluffing is a long-term strategy, and it is an essential element in every successful long-term strategy.

The World of Poker

Poker is a typical example of a competition accompanied by long-term changing fortunes. Without bluffs, poker would not be much of a game. Not only is it terribly boring to play poker with people who never bluff, but in the long run, those who never bluff can only lose. If somebody always behaves exactly as justified by his cards, his opponents will soon see through him. He will win only a little with good cards and will lose exactly as much as is justified by bad cards.

Poker is worth studying because it models many real-life situations: Everybody has lucky and unlucky streaks, but long-term results hardly depend on the luck of the draw. Every accomplished poker player knows that one does not lose too much from a bad hand. The loss is greatest when we have a good hand—only our opponent has an even better one, but we thought he was bluffing. We did not believe him because by his previous bluffs he sowed doubt in us, and we went further than our good cards justified.

Bluffing is like vitamins: It is essential in small amounts, but harmful if used excessively. A player who bluffs too much will be unmasked too often, which would not matter in itself. The problem is that he has invested too much in his later profit, and in the long run he will have a deficit.

How much bluffing is useful? When does it become harmful? Such questions are usually addressed in two ways: with the basically qualitative methods of the philosophical sciences and the basically quantitative methods of the exact sciences. It is to von Neumann's credit that we have quantitative, mathematical methods to deal with such questions. These methods have been of great help in qualitative investigations, too.

A Simple Poker Model

I created the following game for the very purpose of illustrating my point. It is simple enough to demonstrate complex mathematical reasoning in an easily understandable form, yet it is challenging enough to be played in real life. (We have tried it: It is really an enjoyable game, although it cannot rival the depth and complexity of real poker.)

There are two players in the game, X and Y. X is the challenger; Y is challenged. In real life the players can interchange their roles, but we will not do so in our analysis: X will always be the challenger. Here's the game: X rolls a normal, six-sided die, and if it shows a six, he wins, and if it is not a six, he loses. Well, the game is not quite as simple as that. The rules of the game are as follows:

1. At the beginning of each round X puts, say, $10 on the table, while Y puts down $30.
2. Then X casts the die so that Y cannot see the result.
3. Having looked at the result, X does one of two things: He can fold, or he can raise. If X folds, then Y takes all the money. In the present case X will lose $10. If X chooses to raise, he has to put an additional $50 on the table.
4. If X raises, Y can do one of two things: Either he folds, or he doesn't. If he folds, then X takes all the money. In the present case X wins $30. If Y does not fold, he also must put down $50. In this case X has put down $60, while Y has put down $80.

5. If X raised and Y did not fold, then X must show the die. If it shows a six, then X gets all the money, if X was bluffing, then the money goes to Y.

It can be seen at once that the game is almost solely about bluffing. If X never acts as if he rolled a six when he actually did not (X never bluffs), then he will certainly have a deficit in the long run, because Y will eventually always believe him. Thus, when X wins, he wins $30, and when he loses, he loses $10. Over the long term he will roll a six in one time out of six, and so on average, he will lose five times $10 for each $30 he wins. Thus, if he never bluffs, he will lose $50 for every $30 he wins.

If X does not bluff well—if Y can tell when he doesn't have a six—then he will be even worse off. For when X's bluff succeeds and Y folds, X wins $30, but if Y calls his bluff, X will lose $60.

If X bluffs a lot, then he will not profit much even if he bluffs well, that is, if he keeps a poker face. For instance, if X always raises, then Y will probably never believe him after a while and will never fold. Thus, in every six rounds Y will win five times $60, or $300, while X will win $80 only once.

It seems that X must have a deficit in this game no matter what he does. Nonetheless, I would gladly undertake to play the role of X for a length of time. Not because I have a special talent for bluffing. Nor because I would trust my acting ability. Nor would I undertake the role of X on the basis of my knowledge of human psychology. On a purely mathematical basis, X will always win in the long run.

If I had to make a living at this game, this is how I would play. I would memorize a long section from the most boring book in the world, perhaps a table of random digits, leaving out zeros. Then my playing strategy would be as follows: Whenever I throw a six, I raise. Whenever I throw anything else, I raise only if the next number in the table is nine. I am certain to win with this strategy in the long run. Thus, I would not simulate emotion but would be as inexpressive as possible and trust the timing of my bluffing to chance. In other words, I would use a mixed strategy such that in case of a non-six I would raise with a probability of one in nine. We shall soon see why I would bluff with a probability of precisely one in nine and what this strategy is good for. But first let us examine what kind of behavior is required of me.

The distinction between dissimulation and being inexpressive is important, because this is what distinguishes bluffing from other forms of lying. A liar says something that is not true, and in order to be credible, he is forced to dissemble in order to give artistic verisimilitude to an otherwise bald and unconvincing narrative. A bluffer says nothing that is not true. He does not have to dissimulate. It is sufficient that he remain completely inexpressive, which is not so easy. In this game, no facts are explicitly falsified, and thus there are no lies. Sometimes X raises. Occasionally, his raise is a bluff.

The concept of bluffing, however, is not as simple as that. It would not change our game if instead of saying "I raise," X had to say "I rolled a six." Such a rule would require real falsification, to wit, a lie. Nonetheless, the game itself would not change. We therefore have to relax somewhat the meaning of *bluff* to avoid inconsistency. A barefaced lie can be a bluff if it has no misleading informational content. In our game, in the given context, the sentence "I rolled a six" gives no more information than "I raise." The two texts have the same meaning for Y, namely, that he is challenged, and it is his turn to accept or decline.

The Evolution of the Poker Face

Bluffing appears in the posing fights of animals, as mentioned in Chapter 1. It is the fundamental interest of the posing animal to appear as if it were prepared to pose for an infinite length of time, or at least for an unpredictably long period. Were the animal to show a lack of resolution by the slightest twitch, it would immediately be at a disadvantage. Noticing the twitch, the opponent would be able to last a little longer, even if its own strategy would have caused it to give up the fight in the next moment. Natural selection quickly retaliates against those who do not bluff well. Thus, natural selection encourages the development of a poker face.

Why does the poker face evolve this way, and not in an infinite spiral of artful lies? Because unlike a poker face, real lies have informational content. Something with no informational content is not a real lie. Thus one may draw conclusions from a lie, and natural selection would soon develop a species that could draw important conclusions even from lies and could utilize them, for example, to gain an advantage in posing fights.

Other methods of displaying no information can develop. Baring the teeth and many other movements belong to the ceremony of posing fights. There are ingenious poker players who do not show a poker face. Their faces are quite lively, continuously smirking, kidding the opponent, making true and false statements. But it is very difficult to behave thus and give away no information. The poker face is simpler—at least in principle—and just as effective.

The Poker Model Analyzed

The mixed strategy for our model poker game was as follows: Whenever X throws a six, he raises; otherwise, he bluffs with a probability of one in nine. No more, no less. This seems to be a rather small percentage of bluffing. Experiments show that most people bluff more often and as a result run a deficit in the role of X. Nevertheless, I maintain that the one-in-nine proportion of bluffing brings a modest, but certain, long-term profit. Let us examine the balance sheet in a 54-round game with this strategy. (The number of rounds has been chosen to make calculation easier.)

Let us first calculate how much X is expected to win or lose in the 54 rounds with the above strategy if Y accepts all the challenges and again if he declines all of them.

X is expected to roll a six 9 times out of the possible 54. If Y accepts all the challenges, then X will win $80 in each case, bringing in a profit of $9 \times 80 = \$720$. In one-ninth of the remaining 45 rounds, when he does not roll a six, he will bluff. In each of these five rounds X will lose $60 if Y accepts these challenges. This is a $5 \times 60 = \$300$ deficit. X declines the rest of the rounds, losing a further $400. Thus, the balance after 54 rounds is a net gain of $20 for X if Y accepts every challenge.

If Y folds in every round, then X's 9 sixes will yield him a profit of $9 \times 30 = \$270$. With the subsequent 5 bluffs X wins another $5 \times 30 = \$150$. Giving up the remaining 40 rounds (perhaps unnecessarily, but let's stick to the given strategy) will yield a $400 deficit. Thus, X's balance at the end of the 54 rounds is again $20.

Thus, if X plays the above strategy, it does not matter to him whether Y accepts the challenge or declines it. In either case he will win on average $20 in every 54 rounds. Long-term profit can be ensured independently of Y's response, provided that Y has no additional information; that is, X must maintain a perfect poker face.

Can X increase his profit by changing his ratio of bluffing? If we repeat our calculations for an arbitrary ratio of bluffing, then it turns out that if X increases the proportion of his bluffing, he will increase his profit when Y believes his bluffs (and declines his challenges) but will decrease his profit, or even suffer a loss, if Y calls his bluffs. For instance, if the proportion of bluffing is 1/4 and Y always believes the bluffs, the long-term profit in the 54 rounds rises from $20 to $270. If, however, Y never believes the bluffs, then X may expect a deficit of $292.50. Thus by increasing the ratio of bluffing, X gives up a small but certain profit and lays himself open to risk. If Y is stupid and continually falls for X's bluffs, then X can win big. However, if Y is clever, X will soon find himself in the hole.

We can also calculate what happens if X bluffs less often. Then X is once again at the mercy of Y. X can win a lot if Y is not inclined to believe his bluffs, but he will ultimately lose if Y consistently believes the bluffs and declines to challenge, never risking an additional $50.

Thus, the 1/9 ratio of bluffing can be considered as a kind of *equilibrium value* for X. He is assured in the long run of a modest profit, independent of Y's actions. If X is content with his $20 profit, he should stick with this strategy. If it is not enough, he will have to skate over thinner ice by increasing or decreasing his ratio of bluffing according to his judgment of how Y behaves.

We have examined our game so far only from the aspect of X. Now, let's have a look at the game as Y sees it.

We already know that if X is satisfied with his $20 profit, Y can do nothing to prevent a long-term loss. But if X is more ambitious and plays another strategy, then Y has an opportunity to win, or perhaps to lose even more. Based on the previous calculations, if X bluffs more than in 1/9 of the cases, Y can maximize his profit by taking up every challenge—by calling every bluff. If X bluffs less than 1/9 of the time, Y can maximize his profit by declining every challenge.

Real-life situations are not so simple. A certain fluctuation is likely to develop between the two players. As X increases his bluffs, Y in turn accepts the challenges increasingly often. X then responds by bluffing less frequently, and as a consequence, Y declines the challenges more often. Y may also take the initiative, and then X may adapt his play to the level of Y's credulousness. From the psychological viewpoint, this struggle for dominance is the essence of such games, but let us remain with the mathematical aspects of the game for the time being.

If Y knows the precise mathematical workings of the game and is willing to lose $20 in every 54 rounds but does not want to risk more, he may do the following. He can decide to accept every challenge with a probability of 4/9. Why 4/9? We shall soon see.

Let us assume that X always raises (X bluffs). In this case, in 54 rounds X expects to win four times $80 from his expected 9 sixes (because Y accepts four challenges out of the nine) and to win $30 in each of the five remaining rounds (because Y will fold). This is $4 \times 80 + 5 \times 30 = \$470$ profit, thus far. X, who is always bluffing, will win 5/9 of the time with his remaining 45 bluffs; that is, he will win $25 \times 30 = \$750$ (because Y folds in 5/9 of the cases). In the remaining twenty bluffs (when Y does not fold) X will lose $60, which amounts to $20 \times 60 = \$1200$. Thus, X's net balance at the end of 54 rounds of always bluffing will be +$20.

If X never raises when he doesn't roll a six (X never bluffs), then he will win $470 with his 9 sixes (as before) and will lose $10 in each of the remaining 45 rounds, for a 54 round balance of +$20.

Thus, if Y decides to accept every challenge with a probability of 4/9, then it makes no difference to him whether or not X bluffs. Y expects a deficit of $20, no more, no less, in every 54 rounds.

It may seem a small wonder of numerology that the numbers magically saw to it that without any psychology, X can on average win $20 in every 54 rounds no matter what Y does, but Y in turn can manage on average to lose only $20 in every 54 rounds no matter what X does. Once X chooses the strategy of bluffing with probability 1/9 whenever he does not roll a six and is able to show a perfect poker face, Y may twist and turn and maneuver as he will, but he will not win. Y cannot escape losing $20 in every 54 rounds, on average. Once Y agrees to accept every bluff with a probability of 4/9, then X may twist and turn and maneuver as *he* will, but he will not, on average, win more than $20 in every 54 rounds.

These strategies for X and Y make possible a *stable equilibrium*. Once one of the players adopts the conservative strategy, the wisest thing the other player can do is to adopt the corresponding strategy. He cannot come off better, only worse. The recognition of this equilibrium is the essence of game theory.

Bluff Big or Not at All

One of the benefits of game theory to economists is that they can use it to tell whether a game (market situation, economic regulations, etc.) has become unjust for one of the participants; for example, if it will cause the bankruptcy of one of the participants. Thus it turned out that our simplified poker, although at first it seemed to be disadvantageous to X, is in fact detrimental to Y. Thus this game is theoretically unjust, and with sufficiently adaptive players (or those who are knowledgeable about game theory), this would sooner or later become evident in practice. However, the situation can be rectified by changing the rules slightly.

If, for instance, the amount to be paid upon raising is decreased from $50 to $40, the game becomes just. The mixed strategies leading to equilibrium also change. For X, the mixed strategy leading to equilibrium is now to bluff with probability 1/10 whenever he does not roll a six. For Y, the strategy is to accept every challenge with probability 1/2. The expected profit of both players is now zero in the long run. If we further decrease the amount to be paid upon raising to $30, then X is now expected to lose $2 in every 7 rounds in the case of the mixed strategy leading to equilibrium. Similar mathematical tools provide economists with a valuable technique for fine-tuning their economic models.

It is clear from the above calculation that the greater the potential profit in making a bluff, the more often it is worth bluffing, although even when gambling for very high stakes, X will be bluffing relatively infrequently. After a short calculation (not to be detailed here), we would find that if the rules of the game allow X to raise by any amount between $10 and $50, it is worth bluffing only with $50, never less. Every smaller bluff decreases X's expected long-term profit. The moral of the story is this: Bluff big or not at all.

Bluff as a Cognitive Strategy

In mathematical game theory one does not speak of bluffs, but only of different possible moves. The diversity of possible moves includes the diversity of conditions as well, such as whether X throws a six. Mixed strategies are reflected in the probabilities associated with possible moves. It is not mathematical terminology to call certain possible moves *bluffs*. In mixed strategies ending in equilibrium, if the probabilities associated to the possible moves that are regarded as bluffing are not all zero, then *an equilibrium can be reached only if the players are occasionally bluffing*.

It is in the nature of things that every new enterprise is launched with a certain degree of bluff. A new business finds it useful to give the impression of being a long-established enterprise. Otherwise, it would not be able to compete effectively for clients. A contractor does not necessarily have to lie—falsifying the year the firm was founded or predating its previous accounts. It suffices for it to imitate its older successful competitors. On the other hand, it could be beneficial if the contractor openly admits the youthfulness of his enterprise and that he is still unsure in many things. While such a strategy may make it difficult to charge high prices, it is good for attracting understanding clients, giving the new firm a stable customer base. With knowledge of game theory, here, too, a mixed strategy is probably the most effective.

In real life, one does not realize mixed strategies by memorizing tables of random numbers, not even when we really use some kind of mixed strategy. We usually listen to our intuition, which tells us whether to bluff. It has been observed, however, that skilled poker players—in the long run—bluff with the probability associated with the optimal mixed strategy—the one ending in equilibrium. (All the computers in the world working in concert would be unable to calculate the exact optimal mixed strategy for poker, but several approximations have been made.) It is as if the talented poker player had an internal sensory organ controlling the frequency of bluffing, although it is highly unlikely that many poker players have heard of the concept of optimal mixed strategy. Perhaps the acquisition of such unconscious knowledge is expressed by the saying "practice makes perfect." It has long been known that the thinking of the greatest masters of various games and professions is not totally rational but is in large measure intuitive.

A programmed computer would be able to play our little poker game—the just version—without fear of being beaten in the long run. Bluffing or not, the face of the computer exhibits nothing. It probably has a built-in random-number generator, and the computer will stick with its optimal mixed strategy undeceived by even the most sophisticated facial expressions by means of which we may try to fool it. In the case of real poker, however, the computer is in a more difficult situation, since the exact optimal mixed strategy is unknown, and an experienced poker player is able to sense intuitively the slightest deviation from the optimal ratio, thus winning in the long run. With intuition, the human player can set up a system of concepts more complex than the optimal mixed strategy. While this cannot be calculated exactly, this complex, intuitive system may come closer to the optimal strategy than mathematical methods. It was intuition, after all, that invented the concept of the bluff.

Although game theory does not recognize the term *bluff*, we can nonetheless talk about it in a game-theoretical context. We may simply consider every move a bluff to the extent that it is made with a higher probability than the optimal mixed strategy would dictate. It is outside the scope of game theory to ask what the opponent might think when such moves are made against him, even if to a psychologist that might be the main question. From the psychological point of view, such moves are bluffs. Be that as it may, bluffs must appear in every kind of competition, for no equilibrium can be maintained in the long run without them. Thus bluffs are not necessarily ethically condemnable acts, not even according to the categorical imperative.

Nature's Bluffs

Since the discovery of the optimal mixed strategy, the question of how optimal poker players should play is no longer strictly a psychological one. To deduce the optimal strategy, psychological intuition is not only unnecessary, it is useless. It is enough to roll the dice and obey their dictates. Such was the case in the million dollar game. A roll of the die met the demands of morality prescribed by the categorical imperative. Like the yogi who pushes his heart aside before pushing a needle through his chest, the optimal poker player must put aside all his empathy and psychological intuition before the cards are dealt.

Of course, theory and practice are two different things. In practice, the optimal game strategy can be carried out by a roll of the dice only in simple cases. In the real world, as in real poker, the optimal mixed strategy is so complex that it is impossible to calculate. Furthermore, the number of different games encountered in real life is infinite. Even if, thanks to John von Neumann, we are familiar with the concept of mixed strategies, we have to find other, less directly computational, means for their realization. Sticklebacks, for instance, are not particularly well equipped to approximate optimal mixed strategies of posing fights by rolling dice.

The situation can be compared to our sense of time. We are all able to estimate the passage of time, but nobody can do so perfectly, despite that fact that we all possess suites of neurons that fire on a schedule precise to a thousandth of a second. Using a few dozen such neurons, any do-it-yourselfer could assemble a clock accurate to within a few seconds a year. So why didn't mother nature spare a couple of neurons to develop such a clock, which would have enabled us to keep track of time exactly? Perhaps such a clock would have done us more harm than good by requiring us to have access to our low-level neuronal activity, resulting in counterproductive impairment of our more important higher-level activities. We are better off perceiving time with higher-order structures, even at the cost of loss of precision.

Our do-it-yourselfer could quite easily produce a rather good random-number generator from the same neurons, both in sticklebacks and in man. Nevertheless, our mixed strategies are realized through our higher-order activities—emotions, moods, and conceptual systems—and not by neural random-number generators.

Working out optimal mixed strategies requires considerable effort. Like the yogi who has mastered his body to such an extent that with a little "deception"—pushing his heart aside—he can pierce a long needle through his chest, we must master a complex conceptual system and organize and control our emotional reactions in order to approximate optimal mixed strategies. Instead of a random-number generator, nature has developed the poker face, the concept of bluff, the conscience, which in concert are able to realize optimal mixed strategies. Game theory gives us another, nonpsychological, way of studying these mechanisms.

Nature plays by mixed strategies, evolving organisms that have the ability to survive to reproduce their kind. Some of this ability may involve bluffing—behavior that may seem inappropriate to a situation but is nevertheless necessary to achieve a good approximation to the optimal mixed strategy. Thus the pure rationality of the optimal mixed strategy is realized in us—both in our thinking and in our ethical conduct—by means that in themselves are highly nonrational.

THE SOURCES
OF DIVERSITY

6

John von Neumann's Game Theory

It is a fact of mathematics that often the most rational way
of making a decision is to flip a coin.

The fundamental theorem of John von Neumann's game theory
states that in a broad category of games it is always possible to find
an equilibrium from which neither player should deviate unilaterally.
Such equilibria exist in every two-person game that satisfies the following criteria:

1. The game is *finite* both in that the number of options at each
    move is finite and in that the game always ends in a finite
    number of moves.

2. It is a *zero-sum* game: One player's gain is exactly the other's loss.

3. The game is one of *complete information*: Each player knows precisely all the options available to him and to his opponent, the value of each possible outcome of the game, and his own and his opponent's scales of values. (If the game is zero-sum, these two values are the same. There exist non-zero-sum games with complete information content, but von Neumann's theorem does not encompass them.)

According to von Neumann's theorem, it is possible to find an equilibrium in every such game from which it is not worth deviating unilaterally for either of the players, because neither of them can increase their gain this way. The equilibria for these games can always be achieved by mixed strategies. In mathematical terms, this involves finding a *saddle point*.

The Case of the Schizophrenic Snail

To understand the idea of a *saddle point*, imagine the following absurd game. A snail is crawling along a horse's saddle. Now, many snails are hermaphrodites, which for most snails causes no difficulties; but this particular snail, alas, has developed a split personality. The snail's two personalities, let's call them Ferdinand and Isabella, do not get along. They find themselves constantly competing against each other. Ferdinand's paranoia compels him to restrict his movement to the direction of the horse's back—from head to tail—and furthermore, since he suffers from acrophobia, his aim is to get as low as possible in that direction. Isabella, on the other hand, is capable of moving only in the perpendicular direction—from flank to flank. She is a bathophobe—afraid of depths—and consequently, her aim is to get as high as possible in the lateral direction.

If together Ferdinand and Isabella manage to get themselves to the exact middle of the saddle, then they can both relax, for if either of them begins to move, the personality that moves worsens its own position. Thus, the snail's two personalities have come into equilibrium.

Unfortunately, not every surface is saddle-shaped. If our schizophrenic snail were placed on a hilly, uneven surface, at least one of its

selves would always be dissatisfied. At the top of a hill, Ferdinand will move downward, against Isabella's wishes, while at the bottom of a hill Isabella will begin to climb, sending Ferdinand into acrophobic panic. Each will compete for its own interest, and the poor snail will get no rest.

Ferdinand and Isabella are each fully aware of their possibilities. They can move only in the direction allotted to them. Each knows exactly the position of the snail and how much each gains from a particular move. Thus, the game is one of complete information. The sum of the game is also zero, since a millimeter in height gained by Isabella is a millimeter lost by Ferdinand, and vice versa. However, the game is not necessarily finite. It is possible that the snail is pulled about by its two selves till doomsday and beyond, especially on a surface without a saddle point on which they could mutually agree. If the game can be infinite, then von Neumann's theorem does not apply. If, however, the game is constrained by the rule that the snail can make only, let's say, a hundred moves, then the game becomes finite.

In the case of a saddle-shaped surface it is likely that both parties will move toward the saddle point, since it is to the advantage of both to do so. If Isabella moves toward the saddle point and Ferdinand does not, then Ferdinand is only harming himself. If each plays a pure strategy, independent of the other, then each will move toward the saddle point. The expedience of this strategy is so clear that we feel certain that our snail will head directly toward the point of equilibrium, even though it is probably impossible to carry out a real experiment on schizophrenic snails—the relevant psychological tests not yet having been developed. Organisms with even less brainpower than half a snail easily find the shortest path to their goal.

On a surface without a saddle point, however, the favored direction for one personality will usually depend on the direction taken by the other. Knowing where Ferdinand is heading, Isabella will know how she should move, but if Ferdinand knows where Isabella will move, he might choose another direction altogether, one that he finds more suitable. Alas, there is no point on which Ferdinand and Isabella can agree. At any point at least one of them can head in a direction that represents a better bargain. On such surfaces pure strategies are unsafe. The player who outwits the other will win—unless the other uses a mixed strategy.

Von Neumann's theorem states that even if a surface is hilly, uneven, or has no saddle point there is a mixed strategy that leads to a point of equilibrium. This means that if mixed strategies are taken into account, then *theoretically*, the game will become as simple and unambiguous as if the surface itself were saddle-shaped. With mixed strategies, the interests of both parties dictate movement toward this hypothetical saddle point. If one of the players actually moves in that direction, then the other's best move is also to move toward this hypothetical point. The player who fails to do this will suffer a loss.

According to von Neumann's theorem, a balance between the two halves of the snail can develop not only on saddle-shaped surfaces, but even on a hilly, uneven surface, and after a while the snail can relax. The only requirement is that at least one of the selves should know about mixed strategies and should play according to a mixed strategy that suits its own purpose. There is no better strategy. And in addition to reaching the best achievable result, von Neumann's theorem also guarantees this player's security. Even if the opponent plays completely irrationally, such play will not affect the outcome, no matter what ingenious moves it might think up.

Our example of the snail may seem a bit too psychological. Opposing mental forces operate within the snail's dual psyche, each striving to realize its own aims. Yet for the snail as a whole, for its own peace and security, it is important that the opposing forces within him/her find a balance, a position of rest, so that the snail might live in peace with her/himself. For now, our schizophrenic snail will serve to illustrate the ideas of game theory, though as we shall see later, the psychological aspects of the situation are not totally without interest.

The Mathematical Background of von Neumann's Theorem

The proof of von Neumann's theorem begins with the assertion that the expected outcome of applying a mixed strategy corresponds precisely to a multidimensional geometric surface. This was the profound insight that led von Neumann to conceive of games as uniform mathematical structures—despite their complex and chaotic superficial differences—and to study such structures by purely mathematical methods.

This multidimensional surface is an abstraction, for it is difficult for us to imagine even a four-dimensional surface. Even a four-dimensional surface has characteristics that defy our perception of space. For instance, in the fourth dimension it is unnecessary to produce left and right shoes, for given two left shoes one can simply turn one of them into a right shoe, just as we can turn a "p" into a "d" by rotating it in two dimensions and an "R" into a Russian "Я" by rotating it through the third dimension. A cardiac surgeon operating in four dimensions does not have to cut open his patient. He simply reaches into the heart, just as we can touch a line sitting inside a two-dimensional closed figure by reaching into it from the third dimension. As we stated above, every finite, zero-sum game of complete information played with mixed strategies can be considered as a multidimensional geometric surface. It turns out that this surface always has a surprisingly special character: It always has a saddle point.

Although we cannot easily picture a multidimensional surface, we are able to understand some of its characteristics through mathematical abstraction. Thus, we know that a multidimensional saddle point has the same characteristics as those of a horse's saddle—a two-dimensional surface embedded in three-dimensional space. The surface slopes downward in both directions along certain lines, and it rises in both directions along other lines. In von Neumann's construct, these directions correspond to the players' pure strategies—one direction for Ferdinand, the other for Isabella.

Everything we said about the rational behavior of those two snail personalities with regard to the three-dimensional saddle is equally true of abstract, multidimensional saddle surfaces created by mixed strategies. The rational snail personality must approach the saddle point by the methods that are available to it. Only the means of approach have changed. The snail-self no longer applies only pure strategies; it can now approximate the saddle point by mixed strategies. Furthermore, the saddle point itself has become a hypothetical one, located in the abstract world of mixed strategies.

However, since mixed strategies can be realized in practice with appropriate dice (perhaps dice with a hundred thousand sides), this abstract saddle point *can* be achieved in a real-life game. In such cases its abstract characteristics take shape in practice. The mixed strategy corresponding to this hypothetical saddle point has all the characteristics of the equilibrium that we know from theory. As shown by theoretical calculations, the mixed strategy will *really* act as a point of equilibrium. It will realize the same stable balance between two opponents as did the ordinary saddle point.

The Principle of Rationality

In every finite, zero-sum game there is a final outcome, and we can calculate how much one player gained and the other lost. Every player wants his gain to be as large as possible.

The game, then, is about each player seeking to maximize his profit, knowing that his opponent wants to minimize it. To reach their goal, they do not shrink even from the application of mixed strategies. Von Neumann's theorem ensures that in this case each player can achieve his aim, within the constraint of the other achieving *his* aim.

If my opponent is smart enough to minimize my achievement, then the most I can hope for is to reach the saddle point, for my opponent will certainly counteract any moves of mine by a strategy that leads to equilibrium. At the same time, if my opponent assumes that I am also sufficiently clever, then he, too, cannot hope for more. The *principle of rationality* states that we each know that our opponent can be just as smart as we are and that each of us wants to profit as much as possible *on the assumption that our opponent plays optimally*. This means that we do not count on our opponent making a mistake. Von Neumann's theorem expresses that *the principle of rationality can be realized using mixed strategies leading to equilibrium*. Thus, this principle is not simply an attractive utopia, but a practical possibility.

The foundation of von Neumann's theory of games is the principle of rationality. Mathematical game theory assumes that every player acts according to this principle, and for this reason mixed strategies leading to an equilibrium (to the saddle point) are called *optimal*. In Chapter 2, mixed strategies conforming to another principle were called optimal, namely, those leading to a common optimum that might not be in any sense an equilibrium. The difference between these two principles will be touched upon at the end of this chapter.

Our example with the snail was also based on the principle of rationality, the essence of which might also be expressed as follows: Every finite two-person game with complete information can be recast as a struggle between the two selves of a snail. The only difference among such games is the nature of the battlefield on which Ferdinand and Isabella compete and the kinds of moves available to each.

Rational Players

On an actual saddle-shaped surface, the rational strategy of the two players was very simple, and we could expect even a schizophrenic snail to see this and act accordingly.

When the snail moves to more complex surfaces, however, the situation becomes less obvious. With the help of game theory, we humans are able to calculate the probabilities of moves expected of a snail that is clever enough to play an optimal mixed strategy. Thus, at least in principle, we can design an experiment to test the theory. Whether Ferdinand and Isabella, with the limited intellectual capacity of a snail, know anything about such concepts as "optimal mixed strategy," "multidimensional surface," and "saddle point" is moot. We shall discuss this question in detail later, but it is a question in which *mathematical* game theory takes absolutely no interest.

Our understanding of the physical universe took a giant leap forward when Isaac Newton hypothesized totally unrealistic objects that were merely points, having mass but no dimension. With this simplified model he derived a mathematical formula that describes how these objects attract one another under the law of universal gravitation. This abstract model has proved so successful that we can predict the return of a comet or send a spacecraft on target to Jupiter. Analogously, game theory was born when von Neumann hypothesized totally unrealistic, perfectly rational players who are able to think in terms of mixed strategy and to make calculations in complex, multidimensional spaces. There are indications that this abstract model is so successful that with it individual and social conflicts and other decision-making situations can be described, analyzed, and solved. For example, it is through mathematical game theory that it has become apparent that the strange, seemingly irrational strategies of excellent poker players, their second- and third-order actions and bluffs, not only are successful in practice, but are completely rational as well.

No perfectly rational player exists in reality, just as there are no perfectly dimensionless objects or geometrically perfect straight lines to be found in the real world. This fact did not keep von Neumann from building his theory of games, Newton from laying the foundations of classical physics, or Euclid from constructing his geometry. It was others who later examined how well the theory is able to describe the far from perfect objects of the phenomenal world. One cannot expect the theory to work perfectly; but if it can describe reality better than all previous theories, then the new theory will become part of our culture.

Game theory has proved itself, for example, in the sense that unbeatable computer programs can be written for games whose optimal mixed strategies can be calculated. For instance, we can program a computer to play the simplified poker game described in the previous chapter. Programs playing real poker are not yet unbeatable, but the best poker programs, those based on mathematical game theory, are very difficult to beat—average poker players almost always lose against them.

The idea of the perfectly rational player has proved fruitful technically, and the related conceptual system has entered different fields of science, though it can be applied to real people, who fall short of perfect rationality, only to a very limited extent. This can be seen from the fact that most human poker players lose against the best programs and that people usually lose in the role of X in the simplified poker game. As game theory has matured, one may ask why such useful and sophisticated concepts can be applied to people only to a limited extent. This question is related to the fact observed in the dollar auction and prisoner's dilemma that often more rational modes of playing are to be found in the animal world than in human society.

The Value of the Game

In the case of a real saddle-shaped surface it is easy to tell the height of the center of the saddle, and thus a simple measurement tells us the height where our schizophrenic snail will peacefully rest in equilibrium. The analogous height of a saddle point for mixed strategies can be found almost as easily. However, these values are less certain, since where the snail will come to a standstill if both snail-selves play a mixed strategy is subject to chance. In every concrete game, the resting place will depend on what the dice show in the individual moves. However, the *expected* height of the equilibrium point in cases of *certain* mixed strategies *can* be calculated. In other words, we can calculate the average height of the resting place for a game played for many rounds on the same surface.

The expected height of the saddle point is called the *value of the game*. This is the profit (or loss) that a player can secure by playing an optimal mixed strategy. In the case of the simplified poker game analyzed in Chapter 5, the value of the game was $20 in every 54 rounds (as seen by player X), that is, $20/54 = 37 cents per round.

We also learn from calculating the value of the game whether the game can be considered fair. The game is *fair* if its value is zero, that is, if each player can manage, by playing an optimal mixed strategy, not to lose in the long run. Our simplified poker became fair when X was allowed to raise the bet by only $40.

The value of the game shows how much can be won when the opponent, too, gets his turn at bat. According to the principle of rationality, our opponent is fully able to utilize his opportunities. We expect him to do so, and for this we are fully prepared.

If the situation of the two players is completely symmetric, then the game is fair from the outset. If one player can profit in the long run by some mixed strategy, then the other player could do the same by the same strategy. Von Neumann's theorem guarantees that in these cases both players can ensure that their final balances are at worst zero.

Stone–Paper–Scissors

The children's game stone–paper–scissors is a good example of a symmetric two-person game. In this game, two children can show one of three items by a hand gesture: A fist represents a stone, an open hand a piece of paper, and extending the index and middle fingers in a "V" shape a pair of scissors. The stone blunts (defeats) the scissors, the paper wraps (defeats) the stone, and the scissors cuts (defeats) the paper. If both players show the same item, the round ends in a tie. The possible outcomes of the game are shown in the table below, as seen by one of the players. (It is unnecessary to write two numbers in every cell of the table, since the sum of the game being zero, each player's score is the negative of the other's.)

| | | my opponent | | |
		stone	paper	scissors
	stone	0	−1	1
me	paper	1	0	−1
	scissors	−1	1	0

No pure strategy is optimal in this game, that is, none leads to equilibrium, since no matter what pure strategy we play, our opponent can find another pure strategy that defeats ours. For example, if I always show stone, my opponent will sooner or later show paper consistently. If I respond by showing scissors, he will soon switch to stone. If my opponent can see through me better than I can see through him, I'll lose. We know from von Neumann's theorem, however, that there exists a mixed strategy by which I do not lose in the long run, since the game is symmetric. We do not have to perform complicated calculations to find this mixed strategy, since in this case it is very simple:

I show paper with probability 1/3.

I show stone with probability 1/3.

I show scissors with probability 1/3.

If I play by this strategy, then regardless of what my opponent shows, I win with probability 1/3, I lose with probability 1/3, and the round will end in a tie with probability 1/3. My opponent cannot possibly gain an advantage over me psychologically—he cannot see through my strategy, because there is nothing there to see. This mixed strategy ensures that I will not lose in the long run, and since I can expect no better against a completely rational opponent, this is my optimal mixed strategy according to the principle of rationality.

Now, imagine that we alter the rules of the game slightly. My paper, say, is more valuable than my opponent's paper (mine is old parchment, his, cheap newsprint). Thus, if his scissors cuts my paper, I now lose two units. Now the table of the game will look like this:

| | | *my opponent* | | |
		stone	paper	scissors
	stone	0	−1	1
me	paper	1	0	−2
	scissors	−1	1	0

Clearly, this game is unfair for me—the question is, *How* unfair? My first idea is never to show paper, since it puts me at greater risk. I will show scissors and stone each with probability 50%, which still gives me a mixed strategy. As a response, alas, my opponent will show stone all the time; thus on the average, he will win 1 unit every second round, which comes to a net average gain of 1/2 unit per round. However, if my opponent shows stone consistently, then I might show paper sometimes unexpectedly, and I could thereby win. But that is a very risky strategy, since if I lose, I lose double.

For me, the expected value of the "half scissors–half stone" strategy is –1/2 per round. Do I have a better strategy? If there is such a strategy, then it will have to include paper occasionally, despite the risks. Some more complex calculations are required, but in the end we obtain the following:

My optimal mixed strategy:

I show stone with probability 5/12.

I show paper with probability 3/12.

I show scissors with probability 4/12.

In this case, no matter what strategy my opponent plays (pure, mixed, or other), my long-term balance is reduced to –1/12 unit per round—a big improvement. For example, if my opponent always shows stone, then on average there will be 5 ties out of 12 rounds (because I also show stone), I will win 3 times (because I show paper), and I will lose 4 times, because I show scissors. My balance for the 12 rounds will be –1 unit. If my opponent always shows scissors, then on average I will win 5×1 unit out of 12 rounds because I show stone, I will lose 3×2 units by showing paper, and 4 rounds will end in a tie. Thus, my balance will be again –1 unit. The result is the same if my opponent always shows paper and hence if he shows any combination of the three.

We have enough experience now not to be surprised that the number –1 appears in such different ways. Nevertheless, the form of the optimal mixed strategy is interesting. Despite the great risk of showing paper, I will do so surprisingly often, yet I will lose surprisingly little (only 1/12, as compared to the 1/2 of my initial, paper-avoiding, strategy). Thus, this game is less unfair than it seemed at first. We would scarcely have arrived at this conclusion relying only on our mathematical or psychological intuition without knowledge of game theory.

If my opponent is satisfied with an average profit of 1/12 unit per round, he has to play according to the following strategy to guarantee it:

My opponent's optimal mixed strategy:

He shows stone with probability 5/12.

He shows paper with probability 4/12.

He shows scissors with probability 3/12.

It can be calculated again that no matter what strategy I play, my opponent will win an average of 1 unit in every 12 rounds, and once again, the number 1 emerges in different ways.

The stone–paper–scissors game reveals a great deal about the mechanism of von Neumann's theory of games, but because of its very simplicity, it also hides some important features. Thus, it suffices here that only one of the players use the optimal mixed strategy. No matter what strategy the other plays, the outcome of the game will be the value of the game in the long run. This is also true of the symmetric version, but it is not true in general, namely, for more complex games. In such a game, if one of the players plays the optimal complex strategy and the other does not, then the latter can expect to do worse than if he had played optimally according to the principle of rationality.

What, you may ask, does the stone–paper–scissors game have to do with schizoid snails crawling on saddles? The example of the snail was the more general, for it described *every* finite, zero-sum, two-player game of complete information. It is not easy to find the general principle behind the simple stone–paper–scissors game. Stone–paper–scissors has only one move per round, corresponding to Ferdinand and Isabella each making a single decision. The game surface is described by the table of the game—in the case of stone–paper–scissors, the surface consists of nine points. The positive and negative numbers represent various heights on the surface. If player Ferdinand, say, determines the *x*-coordinate, then Isabella determines the *y*-coordinate. With this single move the game is over. The outcome, namely the position of the snail, can be read from the game table. If we imagine this surface as determined by the nine points, then it is far from being saddle-shaped. It is for this reason that we need a *mixed* strategy.

Generalizations of von Neumann's Theorem

Von Neumann's theorem is valid only for finite, zero-sum, two-player games with complete information. Despite these restrictions, it can be applied to a surprisingly large number of games; and if the only merit of von Neumann's theorem were that it provides a general mathematical framework for dealing with such games, we would still consider it a highly significant mathematical accomplishment. It explains the purely rational nature of mysterious concepts like bluff, for example, and it has enabled computer scientists to design excellent poker programs. This theorem, however, has proved to be widely applicable and generalizable to a number of fields.

Von Neumann's theorem is valid for poker only if there are two players. However, poker is generally played by four. Therefore, in order for von Neumann's theorem to be applicable to real poker, it had to be generalized to several players.

That is a different situation entirely, and we meet problems that are nonexistent in two-person games. If, for instance, three players collaborate to fleece the fourth, they will almost certainly succeed even if they do not otherwise cheat. In games for more than two players there is always the danger of two or three players forming a coalition against one or even more players. The theory of games for several players bifurcates right at the outset. On one branch it is assumed that no special agreement or coalition can exist among players. We assume that the rules of the game or natural law prohibits it. The other branch allows for this contingency and looks for possibilities of equilibrium under these circumstances.

Von Neumann's theorem was successfully generalized to coalition-free games for several players by John Nash. He won the 1994 Nobel Prize in economics mainly for this work. He first had to modify the concept of equilibrium, and today the combinations of pure or mixed strategies that result in all the parties being satisfied by each playing the strategy allotted to him is called *Nash equilibrium*. With such a strategy, no player, after learning the moves of all his opponents, can come up with a more advantageous strategy, so long as his opponents do not change their strategies.

Nash demonstrated that in every coalition-free game for several players there exists a (mixed) strategy that leads to equilibrium. Thus, this is a direct generalization of von Neumann's theorem. The only practical problem is that there can be more than one Nash equilibrium in most games, and the value of these games can be substantially different for the different players. It is possible that if several players took concerted action in such a game, *all* of the players would fare better (but since coalitions are illegal, they cannot do this). In zero-sum, two-person games no such problems arise in cases of equilibrium. Even if there are several saddle points, their values are the same for the players.

In the prisoner's dilemma for two or more players there is only a single point of Nash equilibrium: mutual confession. With such a strategy all of the players will be satisfied, for if one of them changes his strategy unilaterally, he will fare worse. The game of chicken has two Nash equilibria—the first player cooperates and the second competes, or the other way around. They are both Nash equilibria, since if anyone changes his strategy unilaterally, he will be worse off. Naturally, however, it does make a difference to the players at which point of equilibrium they end up. Nash's theorem guarantees the *possibility* of equilibrium, but the game can still have very different *values* for the players. Nonetheless, this theory has proved to be very useful in economics, because it makes possible the analysis of different types of equilibrium, permitting the avoidance of the least desirable cases of equilibrium.

The dollar auction game can be studied by the methods of game theory in many ways. If played by people, it is—theoretically—a game for several persons, but in practice, only two persons compete after the first few bids. The game is one of complete information, but it is a non-zero-sum game, since the winner does not get everything that the loser loses. Thus, von Neumann's theorem cannot guarantee the existence of an equilibrium. Nash equilibrium, however, does exist, but it can be achieved only by mixed strategies. In the case of a Nash equilibrium, the players *together* pay one dollar for a dollar on average; that is, the winner pays about half a dollar. As demonstrated by the experiments in Chapter 1, human players usually do not use a strategy that conforms to Nash equilibrium. Nor do animals that engage in posing fights, in which the loser and the winner generally *each* pay about a dollar for a dollar. This can be explained by another application of game theory, namely, by the concept of evolutionarily stable strategies (to be discussed in Chapter 8).

If cooperation is possible in a multiperson game, the situation becomes even more complicated. Many concepts of equilibrium have been developed for such games, and although all of them have achieved partial success in the analysis of certain economic or political conflict situations, no uniform and general theory of coalition has yet been developed.

The games of economics and politics are seldom games of complete information. Even with the best reconnaissance, it is impossible to see completely into the possibilities of the other's technology and decision-making processes—sometimes even the opponent's values are hidden. Thus we have nothing to rely on when preparing the table of the game. The American–Russian strategic arms limitation talks were games of incomplete information (in a certain period they did not even seem to be finite). J.C. Harsányi analyzed these talks by the methods of game theory, and he also won the Nobel Prize in economics in 1994. Harsányi's basic idea was to assume several Russian and American players with different intentions and equipment, and without knowing how the point of equilibrium of these hypothetical players corresponded with that of their real-life counterparts, he examined the points of equilibrium arising from each pairing. Summarizing the results on the basis of probabilities, he created a game to which the methods of games with complete information could be applied, and his clever treatment of this problem produced rather good predictions about how the parties could be expected to react and what agreements could be reached.

It has turned out that von Neumann's theorem can be generalized even more radically. In biology, in psychology, and in economics, it is not always justified to consider rationality as a general guiding principle. *Game theory remains effective in cases of different concepts of rationality*. But let us remain with von Neumann's original idea for the time being.

Games with Handicap

Max Weber differentiates sharply between two types of rationality, namely, between the rationality of values and the rationality of means. Moral philosophers are primarily concerned with the rationality of values. The golden rule and the categorical imperative fall under this rubric. This type of rationality lies outside the scope of von Neumann's game theory, which assumes that completely rational players are fully aware of their own interests and that they know exactly how advantageous the possible outcomes of the game are for them. The theory is not interested in whether these values are real or imaginary, nor in whether a single choice of values can be considered as rational from any (individual, psychological, ethical) aspect. Game theory deals only with the rationality of the means, that is, with the methods of making decisions.

Nevertheless, the choice of values can radically change the nature of the game. As we saw in Chapter 4, for instance, the choice of values according to the golden rule transformed the difficult conflict of the prisoner's dilemma into a conflict-free situation where the choice was self-evident. The selection of values, however, precedes analysis in game theory. Game theory takes the table of the game as a given, as expressing the values of the different players, and it does not deal with the problem of whether such a choice is due to profound general principles or a passing mood.

The assumption that the opponent is a perfectly rational being who is as capable of fighting for his interests as we are is a basic tenet of game theory. In this case, our strategy will be rational if we take this into consideration. It remains to determine the rational methods of implementing this kind of rationality, if indeed there are general means for this. In cases of finite, zero-sum, two-person games with complete information, von Neumann's theorem has demonstrated that we can arrive at a consistent rationality that can be realized in practice by an optimal mixed strategy.

Nevertheless, it is often inappropriate to assume that our opponent is just as good a player as we are. This assumption is certainly false when we play chess against a much weaker opponent. Between chess players of equal strength, it is considered proper to resign when one has fallen significantly behind in material, unless there is a countervailing positional advantage.

In chess, it is not customary for a stronger player to give his opponent a handicap, but in the game of go, for instance, it is almost compulsory. For a Japanese player it is almost inconceivable to play go with a considerably weaker player without giving a handicap. Otherwise, chances would not be equal! For Europeans, inequality seems natural: Clearly, if one of us is stronger, then the chances will not be equal. However, the Japanese think otherwise. For them, a game is worth playing only if the initial chances of the opponents are equal, in which case the player who puts up the better fight will win.

In chess it is not obvious how such fairness can be achieved, but if two runners compete, for instance, then the faster runner can give just so much advantage to the slower that the outcome of the race is determined only in the last few meters. If the amount of advantage is calculated correctly, then the runner who puts up a better fight, conserves his energy better, and mobilizes his resources more efficiently will be the winner. In the case of chess, it is difficult to determine an appropriate handicap, whereas in go the handicapping system is quite refined.

What can game theory say about games with handicap when we assume at the outset that the two players are not equally strong? Interestingly enough, game theory remains just as valid. In a game in which fair handicaps are given, it is not good tactics for the stronger player to set up a clever trap, hoping that the weaker player will fall into it. If the opponent notices the trap, then the stronger player is likely to lose. The stronger player should play completely rationally, with the slight modification of the system of values that gives preference to a complex situation over a simple one. That is how he can maintain tension as long as possible and provide himself with the greatest chance of overcoming his initial disadvantage through the mistakes of his opponent. Thus, he continues to assume that his opponent plays optimally; that is, he continues to apply the principle of rationality, but he compensates for the handicap by trying to create situations in which it is more difficult for each of them to put the principle of rationality into practice, thereby taking advantage of his superiority as a player. Indeed, very strong go players almost never make an incorrect move, even if the handicap is very large. At worst, they enter into complicated situations that they would otherwise avoid—because of their incalculability—when playing against an equally strong opponent.

Games with handicap illustrate once again that game theory deals exclusively with rationality, and that questions of psychology and ethics fall outside the scope of its investigations. In handicap games it is not wise for the stronger player to diverge from the path indicated by game theory, though it is useful to adapt the system of values a little to the given circumstances.

The Part and the Whole

Game theory has become an important tool in practical decision-making. It gives theoretical support to the practice of most sensible investors in placing some of their capital in high-yielding, but very risky, shares and some in shares whose yield is lower, but safer. This is precisely what an optimal mixed strategy would prescribe. In reality, game theory is a theory of rational decision-making.

However, this is only one side of the coin. Game theory can also be considered as a theory that deals not primarily with the players, but with the game itself. Contractors, say, are interested in how to make the most effective and rational decisions in a given economic situation. The secretary of the treasury, however, is interested in the course of the economy itself, whether it will be in balance or in hopeless fluctuation, whether the balance will be acceptable politically, and if not, how general regulations can be modified to make it more acceptable. Thus, the secretary of the treasury is interested in the game itself, rather than in the individual players.

The clinical psychologist working with our schizophrenic snail is interested in the types of forces that are active in the snail, in the strategies of its two competing selves. However, he is more interested in the snail itself. Can the active forces within the snail arrive at an equilibrium, or will the struggle between its split personalities undermine its mental balance? How can the snail achieve a balance that it finds acceptable?

Two levels of game theory are becoming visible: that of the players and that of the game itself. For the psychologist, these two levels appear as different forces acting within individuals and as the psyche itself. However, what for the psychologist is the game is for the economist the player, namely, individuals, with their complex psyches. For the economist, the game itself is at a higher level, at the level of the economy. This situation is similar to the way branches of natural science are built upon one another. Chemistry is built on physics, biology on chemistry. What for one is the game itself is an elementary building block for the other. Game theory operates differently in different spheres of knowledge, but by making it possible for researchers to understand the dynamics of the players and those of the whole game, it has led to important findings in a variety of fields.

The question of the relations between the part and the whole expresses the eternal dilemma of scholars in diverse fields of inquiry. Game theory has provided a radically new and very powerful tool to deal with this problem.

The theory of games has revealed sources of diversity in nature. It became clear from von Neumann's theorem that equilibrium can develop in certain types of games only by the consistent application of mixed strategies. A universal guiding principle, the principle of rationality, leads to diverse strategies. The true significance of the theory is its applicability to many fields of inquiry. In fact, as we shall see later, the principle of rationality itself can be replaced by other, more general, guiding principles, while the basic tenets of game theory remain operational. If there is competition for scarce resources somewhere, a durable and stable equilibrium can develop only if the players apply mixed strategies, that is, if a diversity of individual behaviors, styles of thinking, and coping strategies appear in the game.

7

Competition for a Common Goal

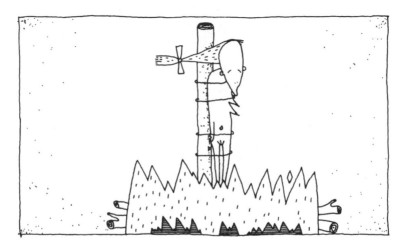

Integrity is unforgivable.

A café ekes out a miserable existence in a small and dusty town in the novel *Kerengő (The Peripatetic)* by the Hungarian writer György Spiró. A stranger arrives in town and opens another café, just across the street from the old one. Everybody thinks him a fool, since the town can scarcely support the old café. Soon, however, both cafés begin to flourish. People now have a choice, and two groups of steady customers form, one for each café, individuals expressing their identities by patronizing one café or the other. Severe conflicts arise between guests of the opposing cafés. Occasionally, a guest passionately withdraws his allegiance from one café and bestows it on the other. Perhaps the only person to frequent both cafés is the novel's protagonist, and even he was not a habitué of the original café when it was the only game in town.

The two cafés are competing, to be sure, but it is not a zero-sum game. When the new café closes, the old one sinks back into oblivion.

In non-zero-sum games the interests of the players are not completely complementary. Almost every human interaction—be it work, conversation, or strife—is a complex mixture of opposing and common interests. Genuine zero-sum games exist only in such manufactured situations as those that take place across the chessboard or around the poker table. Non-zero-sum games are also called *games with mixed motives*.

In games with mixed motives the players can, as usual, win or lose what the opponent loses or wins, but there will also be opportunities for gain or dangers of loss that can be exploited, respectively avoided, by the players only through cooperation. Individual and common interests are thus mixed in such games. Environmental protection is a game with mixed motives. On the one hand, it is in the polluter's interest to spend as little as possible on cleanup, since it produces no direct profit. On the other hand, it is in the common interest to protect the environment.

Von Neumann's theorem deals with zero-sum, that is, *purely competitive*, games. We have seen, however, that his theory has important things to say about other games as well. Let us look at games at the other end of the competitive spectrum: games in which the interests of the two (or more) players *exactly* coincide. Here there is no conflict, and rather than trying to outwit each other, the players strive to cooperate fully.

Purely Cooperative Games

The quiz program *Mr. and Mrs.* that appeared for years on British television is a purely cooperative game. The contestants on this program were married couples, and questions were given to the husband and wife that they had to answer independently of each other. If they gave identical answers, they scored a point, and winning couples won considerable sums of money. The game worked like this: A question was posed to the couple, and they were given a choice of answers. For example: "What kind of flowers would the husband give his wife for her birthday—roses, tulips, carnations, or lilies?" Or "You've decided to redecorate the kitchen. Who chooses the new furniture, the husband, the wife, both together, her mother, or an interior decorator?" The choices were presented to the wife and to the husband in a different order and with variations in phrasing, so as to prevent the possibility of their agreeing beforehand on a system of reply, such as always choosing the third alternative or the alternative beginning with the letter closest to the beginning of the alphabet.

There are many strategies available for playing this simple game—perhaps that is why it became so popular. The worst tactics proved to be for both husband and wife to give their honest opinions. Slightly better was for each to choose the answer they thought the other would pick. The best results were achieved by couples who answered asymmetrically, such as the husband saying what he really thought, the wife what she thought her husband would say.

In purely cooperative games, breaking symmetry often improves the players' chances. When two people lose each other in a crowd, if each searches for the other, they will most likely take longer to find each other than if one of them stops at a busy place and the other looks for him. But stopping is a risky strategy if no agreement was made beforehand. If both decide to stop and wait, they will never find each other! Thus, it is advisable for a parent and child to agree that if they become separated, the child should stop at a visible place and wait to be found.

The jury of a beauty contest also plays a purely cooperative game. Each member of the jury wishes to elect as beauty queen the one most universally considered beautiful, even if had he followed the promptings of his heart, he would have chosen another candidate. What is important is that each judge accept the possibility of his personal candidate losing to the eventual winner. Consequently, the members of the jury will not vote for the woman that they think is the most beautiful. For they all know how individual their tastes are (just like everybody else's), and perhaps others will consider something a defect that for them is a particularly attractive feature. The best strategy is to try to guess which woman public opinion thinks is the most beautiful.

John Maynard Keynes pointed out that professional investors employ such a strategy when they gamble on the stock exchange. They observe public opinion (or rather what passes for public opinion), because the price of a company's shares will ultimately be determined not by the true worth of the company but by its perceived worth. Thus it is that the expectation of inflation—a purely psychological factor seemingly remote from the laws of economics—can generate real inflation. I may think that the economic indicators do not point in the direction of increasing inflation; nevertheless, if I feel that public opinion expects this, then I will buy and sell according to how I expect others to buy and sell. As a result of the expectation of inflation, a kind of cooperation may develop among investors, although perhaps with negative consequences.

A simple experiment may demonstrate how this kind of silent cooperation can develop. Two strangers are asked to participate in the following game. Each of them should say "heads" or "tails," independently of the other. If they give the same response, they get a dollar; otherwise, they get nothing. In Hungary, ninety to ninety-five percent of those responding in such a situation choose "heads." If, however, we observe coin tosses in real life (for example, the choice of side in football), we'll see "heads" only sixty to seventy percent of the time. Thus, most people seem to suspect that the majority says "heads," even those who consider themselves in the minority.

Sociologists are familiar with the phenomenon of a majority thinking that they are in the minority. In an American survey, white parents were asked what they would answer if their daughter asked if her black classmate could come over to play. The choices were as follows:

(a) My child should not play with black children.
(b) My child can play with black children, but only in school.
(c) Why not?

Seventy percent of the parents gave reply (c). However, when the very same parents were asked what they thought the majority of parents would say, only 30% chose "Why not?" Another survey, in Hungary, in 1993, investigated what people thought of a person under 35, a Jew, or a woman becoming Hungary's next prime minister. In each case, the majority responded in the affirmative to all three scenarios, expressing the opinion that what matters is competence, not superficial characteristics. Yet a majority also thought that such people were in the minority. Such a lack of knowledge can make cooperation difficult.

Mutual Fate Control

In life, our fate is often bound to that of another person while we pursue our individual destinies. The Hungarian writer Géza Ottlik writes in his novel *School at the Border:*

> The civilians are also bound together, so are the alpinists in the Himalayas, so are the lovers, because it won't work otherwise. We, however, also knew that beyond all this we played a separate game with our own fate. If Jaks is shot to death at the wall, it will be his business. Any of our sympathy would be mistaken, false and useless, for we cannot even speak of the subject, we cannot tell whether he suffered a final defeat, or just the other way around, he played it low on fate. We did not know his rules, we only knew that Szeredy, and Medve, and me, and Jaks himself, and after a while all of us fought our own great match alone, and that no man on Earth could possibly help. This is the reason why our bond was stronger than that of the lovers or the alpinists, because this knowledge was woven into it from the beginning.

Mutual fate control is a classic experimental situation in social psychology. Two persons—let's call them players—are seated in two rooms. They cannot see each other, and they cannot communicate with each other in any way. There are two buttons in front of each of them, L on the left and R on the right. Neither knows what the two buttons are for, but whenever they hear a high tone, they have to press one of the buttons. After each high tone, they hear a deep tone, after which an "announcement of results" follows: Each player receives either a reward or a punishment. The reward can be money, and punishment can be an unpleasant sound, a slight electric shock, or simply the lack of a reward.

The essence of the game is that in reality, the players send reward or punishment *to each other*. By pressing button R, a player is sending a reward to the other player, while pressing button L sends a punishment. But the players have no idea of this. The question is whether they can develop cooperation, whether they can reach a point at which they thereafter send only rewards to each other.

The players assume that there is a connection between pressing a button and receiving a reward or a punishment, but they don't know what this relationship is. In some of the experiments the players do not even know about the existence of another person; in others they are told only that there is another person in a similar situation in the next room.

Neither player can discover the rule of the game, since they don't have the crucial piece of information, namely, whether the other player won or lost. The player can see only that after pushing a button, sometimes he gets a reward and sometimes he receives a punishment.

Actually, the game is not fully cooperative. A player knows only his own interests and outcomes, and so it can make no difference to him whether he sends punishments or rewards, provided that it is rewards that he receives. However, sending punishments while receiving rewards is an unlikely development, one that never occurs in practice. We win only if we can teach the other what is good for us, if we can develop a kind of cooperation in which we can send rewards to each other. Experiments like this model the development of cooperation in a pure form.

There are many variations on this game, and if we are clever enough, we can hope to reveal rules about the general nature of cooperation.

Theoretical Considerations

One of the laws of psychology, Thorndike's *law of effect,* has been verified repeatedly in human and animal experiments in a variety of experimental situations. If we are missing essential information about the rules of a game, it seems logical to adhere to the following principle. If our action is followed by something good, we should repeat this action, if our action is associated with something bad, we should try something else. In short, do not abandon a winning strategy, but throw a losing one overboard. As Thorndike put it in his often cited book *Animal Intelligence: Experimental Studies*, published in 1911,

Of several responses made to the same situation, those which are accompanied or closely followed by satisfaction [are] more firmly connected with the situation…those which are accompanied or closely followed by discomfort…have their connections with the situation weakened.

This phrasing is more refined than the summary principle "Do not abandon a winning strategy"—and we'll soon see why. Let's examine what this strategy predicts. Due to its extreme simplicity, we shall call this strategy the *polarized law of effect*. At first it seems unlikely that there is anything better, since we do not know the rules of the game. Thorndike's law is usually valid in situations where the subject evokes reward or punishment by his own response. Similar results obtain in the mutual fate control experiments described above.

The polarized law of effect is very similar to the *tit-for-tat* strategy. If we knew when we were sending reward and when punishment (or even the fact that we are sending something to our partner), the two would be exactly the same.

In our mutual fate control situation, when the players send reward or punishment to each other, it is easy to follow the operation of the polarized law of effect mentally. At first it is a matter of chance which buttons the players press. If they both press R, sending reward, then both are rewarded immediately, and according to the polarized law of effect, they will not change their strategies, and thus cooperation develops immediately. This will be depicted as follows:

Player 1: $R + R + R + \cdots$
$$\times \quad \times \quad \times$$
Player 2: $R + R + R + \cdots$

The plus sign "+" means here that the player received a reward, while the arrows indicate that this effect was due to the button pressing of the other player. Punishment will be indicated by a minus "–" sign. If both players begin the game by pressing button L, the game will take the following course:

Player 1: $L - R + R + \cdots$
$$\times \quad \times \quad \times$$
Player 2: $L - R + R + \cdots$

If one of the players starts with R and the other with L, then the game will proceed thus:

Player 1: $L + L - R + R + \cdots$

Player 2: $R - L - R + R + \cdots$

Since there are no further possibilities, we can say that the polarized law of effect predicts that cooperation will develop after the third round at the latest and that it will be maintained.

Experimental Evidence

In order to test the above hypothesis, precisely designed and controlled experiments were conducted by H.H. Kelley and his colleagues. The predictions of the theory were only partially confirmed by the experiments. Stable cooperation never developed after the third button pressing. At the beginning, all of the subjects pressed the buttons at random, presumably trying to figure out the relationship between the button pressed and the outcome. On average, in series of 100 mutual button presses, the subjects pressed button R about 75% of the time. In 96% of the pairs, cooperation had developed by the end of the series. It took an average of about 50 button presses before stable cooperation developed. These human players did not behave according to the law of effect, at least not its polarized version. A significant observation is that the players never guessed the rule of the game, even after stable cooperation had developed.

Why did we say above that the experimental evidence only partially confirmed the predictions of the polarized law of effect, rather than that the evidence disproved the theory? There are two reasons for this. First, in the great majority of the experiments, cooperation did in fact develop, which is in accordance with the theoretical prediction. Nonetheless, the number of button presses needed for cooperation to develop was much greater than the theory predicts.

The second reason is apparent. If we consider the three theoretical possibilities, we see that unless the pairs are able to cooperate immediately, cooperation can develop only after mutual punishment. This prediction was supported perfectly by the experiments. Such precision was unexpected, since in psychological experiments complete consistency is rare. But in the experiments carried out by Kelley and his colleagues, every long, mutual R series was without exception preceded by mutual L pressing, as if the *precondition* for cooperation were mutual mistreatment.

Relying on theoretical considerations and empirical observations, social psychologists have long known that an ugly quarrel or an unpleasant conflict is often necessary before mutual trust or even intimacy can develop. Criminologists are also aware of the fact that trust and cooperation often develop among gang members only after a violent fight. Robin Hood's friendship with Little John began only after their battle on the bridge. Or as Algernon observes in Oscar Wilde's *The Importance of Being Earnest,* women who come to call each other "sister" usually do so only after they have called each other a lot of other things first.

The main virtue of the experiment of Kelley and his colleagues is its abstractness, that the players are not aware of the rules of the game and, further, that they do not know the rules even after they have entered into stable cooperation. We are rarely aware of the motivations for our actions, let alone those of our partners. If the sole result of the theoretical analysis based on the polarized law of effect had been to call the investigators' attention to experimental data worthy of further study, then such theorizing was not useless. This simple theory, however, has indeed led to further results.

Asynchronous Decisions

So far, we have studied mutual fate control in situations where the players always make their decisions and receive their rewards or punishments simultaneously, at the sound of a tone. What happens if the timing of the tones is made asynchronous?

Such a change is an apparently insignificant technical alteration. After all, neither player has any idea what is happening in the other room, no notion of when and in what sequence the tones are presented. Moreover, there is no way of acquiring this information, since each player knows only that after hearing a high tone he has to press a button, and then he hears a deep tone, after which he receives either a reward or a punishment. Then this whole process is repeated. There seems to be no reason why the timing of the tones should make any difference in the outcome of the experiment. Surprisingly, it does.

What does the polarized law of effect say about this case? Since the players are no longer acting in synchrony, the player who presses the button first receives no response on the basis of the other player's move. Therefore, let us assume that the first player receives an artificial positive response (it could as well be negative) to the first push of the button and that the actual mutual fate control begins only subsequently.

If both players press button R in the first trial, cooperation develops immediately, and there is no reason why it should break down. But if one of the players presses button R and the other presses button L, then the course of the game will run like this:

Player 1: $R + R - L - R - L + L + L - R + R - L - R - L + L + L - \cdots$

Player 2: $L + L + L - R + R - L - R - L + L + L - R + R - L - R - \cdots$

After the seventh round the snake bites its tail, and the series will be repeated endlessly. Both players will press the series $R\,R\,L\,R\,L\,L\,L$, although with a shift in time.

The situation will be the same if both players press button L in the first trial. Thus, the theory predicts that no cooperation will develop in an asynchronous situation!

In fact, the theory further predicts that if the players behave according to the polarized law of effect, they will achieve a worse outcome in the long run than if they had pressed the buttons at random. For according to the theory, a reward will be sent less than half the time (3 times out of 7). Even the absurd strategy "change buttons only if you are rewarded" leads to a better outcome than this!

The theory predicts an even worse outcome if the game is timed so that player 1 presses a button, then player 2 receives his reward or punishment, then player 2 presses a button, then player 1 receives his due, and so on. Thus, if player 1 begins with button R and player 2 begins with button L, the course of the game will run like this:

Player 1: $R \quad\;\; -L \quad\;\; +L \quad\;\; -R \quad\;\; -L \quad\;\; +L \quad\;\; -R \cdots$

Player 2: $\quad +L \quad\;\; -R \quad\;\; -L \quad\;\; +L \quad\;\; -R \quad\;\; -L \quad\;\; +L \cdots$

Once again an infinite loop develops, this time even worse. According to the theory, the players will send punishment to each other two-thirds of the time. The situation is the same if both players happen to start with button L.

Thus, theory predicts that no cooperation will develop in an asynchronous situation. In fact, the players will send each other more punishment than reward. Experiments only partially confirmed these predictions. It was found that cooperation did not develop easily. Only 2 out of 50 pairs developed cooperation in 150 cycles of button pressing. Thus, only a small percentage of the pairs managed to develop cooperation in an asynchronous situation compared to the synchronous situation, in which over 96% of the pairs cooperated. The theory proved correct in that it predicted a significant, qualitative difference between the synchronous and asynchronous situations. However, contrary to the prediction of the theory, the proportion of those pressing button R was close to 50% in the long run, which isn't saying much: To achieve 50% it is sufficient to press the buttons at random.

These two results *together* are not self-evident at all; indeed, they are rather surprising, especially if we consider that the players have no knowledge about the game's synchronicity or lack thereof. This is why we said earlier that the experiments only *partially* confirmed the predictions of the theory; this is why we considered it important that despite the inaccurate quantitative predictions, the theory predicted correctly the qualitative fact of the development of cooperation in a synchronous situation.

The Role of Being Informed

Thorndike's law applies to situations where the subjects elicit reward or punishment through their own actions. If the players have no idea about the existence of a partner, they have no a priori reason to believe that reward and punishment depend on something other than their own actions. In such cases Thorndike's theory can be applied in its pure form. If the players know about the existence of another player, then, of course, they might suspect that there are influences beyond their own actions. Yet if the polarized law of effect worked perfectly, it would make no difference at all whether or not the players knew of the existence of their partners. Therefore, Kelley and his colleagues studied whether the outcome of the experiment was affected by the players' knowledge that they had a partner.

Kelley and his colleagues performed three variations on their experiment. In one, the players did not know about the existence of the partner. In the second, the players knew about the existence of a second person but did not now that this person was participating in the same experiment. In the final variation, the players knew that there was a partner in another room, but they were not informed as to the essence of the relationship between them. Some players knew only that there was a partner, while others could communicate with the partner to some extent.

The degree of information about the partner did not alter the above-mentioned tendencies, but it strongly modified the particular outcomes. The more information the players had, the more R-button pressing occurred. This increase was small but significant.

In addition to the law of effect, our actions are also influenced by other forces in situations of mutual fate control. Thorndike already noted that the law of effect can be observed more clearly in animals than in man. Thus sticklebacks follow the tit-for-tat strategy better than human beings, who, being able to utilize information outside the scope of the immediate situation, produce behavior that does not strictly follow the theory. In certain situations, like the dollar auction or prisoner's dilemma, the results, as we have seen, can be positively detrimental. In other cases, however, such as the asynchronous versions of mutual fate control, human capabilities can pay off with interest, as was the case with the two pairs of players who were able to develop cooperation despite the theoretically hopeless situation.

The Rationale of Curfews

Is it really the synchronicity (or asynchronicity) of the situation that determines whether cooperation develops, or is there another, hitherto unremarked, factor at work. What happens if feedback is delayed? What does our model predict for delayed feedback in a synchronous situation? Let us assume that the feedback is based on the penultimate button pushed so far, rather than on the most recent one. In this case, if one of the players starts with button L, the other with R, the game will proceed like this:

Player 1: $L + L + L + L - R + R - L + L + L + L - R + R - L + L \cdots$

Player 2: $R + R - L - R - L - R + R + R - L - R - L - R + R + R \cdots$

Here the initial feedback is artificially positive for both players, since there is nothing yet to feed back. The outcome is the same if we give negative feedback after the first L pressing.

According to the model, in this case not only does cooperation fail to develop (an infinite loop of six turns develops), but the symmetry between the two players is also upset: One of the players presses button R more often than the other. Relying on our results so far, we may say that the lack of symmetry is not a serious problem, since the players experiment with both buttons repeatedly at the beginning of the game, and if the polarized law of effect is not working, they will not play according to this strategy at all. However, the model's prediction that no cooperation will develop has to be taken seriously.

If we can influence the rules of the game—and our aim is to have cooperation develop as soon as possible, with as great a probability as possible—then the decisions of the players should be made more or less simultaneously, and the players should be informed about the effect of their decisions as soon as possible. Military planners know this—intuitively or from historical experience—and so when there is a threat of civil unrest, when it is essential that the predisposition of the populace be revealed as quickly as possible, a curfew is sometimes ordered, and news about it broadcast frequently. The purpose of the curfew is not to shoot, and thereby eliminate, disobedient elements, but to make individual moves (of those in the street) and the evaluation of the moves (by those at home on account of the curfew) as synchronous as possible.

The results also explain why the dollar auction proved to be a greater problem than the iterated prisoner's dilemma. The dollar auction is a typical asynchronous situation, in which the players make their bids by turns. In the iterated prisoner's dilemma the decisions are made in synchrony. This is the reason why the tit-for-tat strategy offers a reassuring solution to the prisoner's dilemma but not to the dollar auction.

On the Nature of Psychological Laws

It is understandable now why Thorndike was so careful in phrasing his famous law. The law of effect, on which we have based our theoretical investigations, evidently fails in its "polarized" form, a finding supported by both "synchronous" and "asynchronous" experiments. Nevertheless, the law says something important about human (and animal) behavior. The evidence of mutual fate control experiments is that even this artificially polarized version of the law of effect makes useful predictions. In fact, it has laid the foundations of experiments whose results are not at all self-evident, that sometimes even go against intuition. Psychological laws are different from physical laws. A physical model that gave such inaccurate quantitative predictions would hardly be taught 80 years after its first enunciation.

Based on the outcomes of the synchronous and asynchronous situations, nature would have little use in creating beings that stuck to the law of effect too rigidly. It may be essential for social organisms to cooperate in a variety of situations. We are often in an asynchronous situation with our partners, and it would be to our disadvantage if the law of effect were implanted in us too inflexibly. In synchronous situations, however, it may be very useful if Thorndike's law is in our strategic armamentarium—and according to experimental evidence, it is there to some extent.

The theory may be quite shaky from the perspective of one educated in the natural sciences, and the exactitude of the predictions is really unsatisfactory. But even the theories of physics are not quite satisfactory for a mathematician. For example, almost every step in the derivation of Schrödinger's famous wave equation is mathematically incorrect (at least the derivations I know). Nonetheless, as we shall see in Chapter 10, Schrödinger's derivation is undoubtedly one of the finest examples of physical intuition. With his equation he laid the foundations of a highly fertile theory and contributed to our more precise knowledge of the universe.

This is true also of Thorndike's law, as it is of other important real-izations in psychology. Evidently, our model says nothing about what may be going on within the players when in the asynchronous situa-tion cooperation finally develops. Nevertheless, intuition based on the general conclusion drawn from the investigation of the synchro-nous situation proved to be correct in the asynchronous situation, too. Both pairs who ultimately reached stable cooperation mutually hurt each other immediately before beginning their cooperation.

The reasoning applied in the analysis of mutual fate control has typically shown how a purely theoretical, mathematical model can lead us to interesting and valid psychological conclusions. Cognitive psychology is based on such methods, which are well established in the other natural sciences. One sets up purely theoretical models, an-alyzes the theoretical consequences of those models, and investigates the validity and limits of the models by experimentation. This is what we shall also do with game theory.

Cooperation by Competition

Whether or not a game is competitive is often determined by circum-stances, and not by the rules of the game. As we saw earlier, the sec-retaries in Merrill Flood's institute were able to treat an apparently competitive situation as a cooperative game. Armed with the golden rule and the categorical imperative, we can transform many compet-itive situations into purely cooperative ones. But the opposite is equally true. Often, competition is the best way of achieving useful cooperation, for example in certain segments of a free-market econ-omy (Chapter 9). But even under much simpler circumstances com-petition can be an effective tool in achieving a common goal.

In order to investigate this issue further, we prepared a short video game called *Rabbit Hunt*. It was a two-person game, and the aim was to catch all the rabbits running in a field. In one version, the two play-ers competed against each other, and the player who caught more rab-bits was the winner. In the course of the game the scores of the play-ers could be seen on the screen. In another version, the aim of the two players was to catch all of the rabbits as quickly as possible. In this version, it was of no account how many rabbits the players caught separately, and only the elapsed time and the number of rabbits caught could be seen on the screen. The first version is purely competitive, the second one purely cooperative.

Two series of competitions were organized, an individual, round-robin competition with the competitive version and a competition for pairs with the cooperative version. The participants of the individual competition did not know that the computer was recording the time required for the two players to catch all the rabbits.

In the competitive game, if a player could not catch a rabbit, it was in his interest to move the rabbit away from his opponent, and this occasionally really happened. In the cooperative game the players had the same options as the players of the competitive game. Thus, in principle, the same players should have achieved better time results in the cooperative version. Nevertheless, on average, the competitive players caught the rabbits in less time than the cooperative players, and the best times were achieved in the competitive games— even though the best and the second-best players in the competitive game repeatedly attempted together to break the record in the cooperative version.

8

Hawks and Doves

An altruistic person does good primarily to his own altruism.

The common wisdom is that natural selection and the struggle for survival are cruel laws of "Nature red in tooth and claw." The stronger and smarter will promote survival of the species; the rest should perish. If such are nature's laws, no wonder they apply to us humans as well. Plautus's observation that "A man is a wolf to another man . . ." expresses this very idea.

This saying is unfair to wolves, however, for with wolves, no matter how cruel the struggle, the winner never rips open the throat of the loser. Knowing the cruel massacres of human history, or even our irrational behavior in the dollar auction, perhaps the human condition is actually worse than what the saying suggests. Perhaps natural selection is not as cruel a mechanism as it first appears, and the root of man's inhumanity to man must be looked for elsewhere.

We know that nature continually produces new types of organisms, which reproduce to create more or less exact copies of themselves. *Evolution* is the resulting change, brought about primarily by natural selection. While natural selection may the primary force at work in evolution, it may not be the only mechanism. We can think about the concept of evolution without reference to the precise mechanisms of natural selection and thus study nature's *modus operandi* at a more abstract and general level.

Analogously, gravity is the name given to the natural force that attracts objects with mass to each other, as a result of which the planets revolve about the sun and we do not fall off the earth. Newton demonstrated that with the concept of gravity as a physical force we can deduce the laws of the movement of bodies with great accuracy. Darwin proved that evolution is an important biological concept by means of which the laws of the development of species can be described. These scientific concepts—since they describe the essence of the thing clearly and effectively—have entered everyday thinking. They form the foundations of the scientific thought even of those who believe in the doctrine of divine creation. Indeed, in 1996, the Pope himself declared that Darwinian evolution can be reconciled with Catholic dogma. Regardless of whether the world was created by God or it simply evolved by itself, the basic principles of how the world operates may include both gravity and evolution.

However, all this does not mean that we really understand these concepts. Indeed, the traditional view of gravity conflicts with more recent evidence of quantum physics.

Jacques Monod has written, "The theory of evolution has a special characteristic: Everybody thinks he understands it!" According to current opinion, the theory of evolution does not explain the *origin* of species at all. Rather, it is a higher-order natural law that describes change and stability in a world in which life already exists. This principle can be applied to the development of species, but this has nothing to do with the abstract theory, just as the theory of gravity is not altered by the fact it can be applied to explain planetary motion in our solar system.

Game theory also deals with questions of stability. Von Neumann's theorem proved this possibility by a clever mechanism, namely, mixed strategies—at least in the case of certain types of games. Heterogeneity follows necessarily from this line of reasoning, strongly echoing the kinds of problems studied by biologists. But the fundamental hypothesis of game theory, namely the principle of rationality, mitigates against the application of game theory to biology. It is difficult to imagine that animals assume that their fellow creatures operate according to purely rational behavior.

Rationality, however, *can* be the basis of an abstract, higher-order guiding principle, namely, evolution. In this case, it is not necessary to endow every organism with the power of flawless logical reasoning. It is sufficient that every form of life that does not conform to this rational principle be sentenced to extinction by natural selection. But there is no general agreement that the principle of rationality can serve as a cornerstone of evolutionary theory. Perhaps there is an alternative, more plausible, general principle.

Our analogy between gravity and evolution was not merely incidental. As we shall see in Chapter 10, certain questions arising in quantum physics are surprisingly similar to those arising in research into the basic principles of evolution.

When looking for the ultimate rational principles that govern the universe, we can easily find ourselves in a situation like that of the man who goes into an army surplus store and inquires, "Do you have camouflage jackets?" to which the salesman replies, "We do indeed, but we have been unable to find them."

The Theory of Group Selection

Initially, the scientists studying evolution thought that natural selection operates at the level of the individual organism. Each individual struggles for existence, natural selection favoring the better adapted: those who produce more offspring who themselves go on to reproduce. Such an approach, however, failed to explain why animals treat one another so leniently, to the extent that there is to be found in the animal kingdom what can only be called altruism. For instance, our friends the sticklebacks are frequently willing to sacrifice themselves for the benefit of the group. In the case of bees, natural selection is even more difficult to interpret, since the majority of the individuals in a bee colony do not participate in reproduction at all, but pass their lives as sterile worker bees, gathering nectar and toiling in the hive to promote the general welfare.

Darwin proposed that perhaps natural selection exerts its effect as though the entire colony of bees were a single organism. Yet he saw the difficulties in this approach. In his diary, he periodically under- lined in red the facts that seemed to contradict his theory. The most persistent problem was the following: Characteristics useful for sur- vival can be passed from one generation to the next only through the reproduction of individuals. Therefore, natural selection can be the exclusive mechanism of evolution only if it favors the individual as well as the group. This hardly seems to be true of self-sacrifice.

According to the hypotheses of the *theory of group selection*, the unit of natural selection is not the individual, but a larger group—a class or even a whole species. If selection really affects a whole species, then it really can force individuals to sacrifice themselves, for then precisely those groups that exhibit altruism will survive. Such altruism need not be pure. Sticklebacks, for example, will sacri- fice themselves when the occasion demands, but they also compete as individuals for food, territory, and mates. This does not contradict group selection theory. As we saw in the rabbit hunt game, competi- tion may also serve the good of the group. Thus, it is conceivable that there are mechanisms by which natural selection affects the whole group through competition among individuals.

The theory of group selection is strikingly different from Darwin's theory, but rather than solving the problems indicated by the red un- derlining in Darwin's diary, it simply erases them. The fact that these problems remain unsolved does not vitiate the entire theory of group selection. Every scientific theory has unsolved problems. If there is evidence that competition can be a tool for cooperation, then the proponents of group selection may have a case. Konrad Lorenz has demonstrated that competition for territory among individuals can serve the interest of the species in that the species then utilizes the available territory optimally by preventing overpopulation.

Despite such evidence, the majority of biologists have not supported the theory. It turns out that the questions posed by Darwin can be answered without assuming that natural selection affects units larger than the individual. Surprisingly, a logical and general solution at least as good assumes that natural selection affects units far smaller than the individual. We shall use wolves to illustrate this paradoxical idea.

When wolves fight, the victorious wolf never kills the vanquished. Why? If two strong wolves fought each other, then the elimination of one of them would result in a weaker pack. It is not in the interest of the species to develop a killer instinct—which to proponents of group selection is self-evident.

The Theory of Gene Selection

Imagine for a moment that natural selection does not affect the whole pack, but only the survival of the instinct that determines whether or not the winner of a fight kills its defeated opponent. Let us assume that in the population there are two genes that govern this instinct. One of them dictates, "Kill your opponent! You will eliminate a rival." The other counsels, "Have mercy. Don't kill him. You have nothing to gain thereby." In reality, of course, genes do not engage in such dramatic dialogue with those that bear them. Neither wolves nor genes have to justify their behavior. A wolf having one type of gene can behave only according to that gene's dictates and can no more act differently than it can have blue eyes if it has genes for brown. Given these two genotypes, which wolf will be fitter for survival? It may seem at first that the killer gene is superior, since every killed opponent means one competitor fewer. However, it may be that by killing an opponent a wolf possessing the killer gene has simply done its rivals a favor. Thus, the wolf with the altruistic gene may be better off in the long run—in which case natural selection will favor this gene.

According to the hypotheses of the *theory of gene selection*, the unit of natural selection is not the individual, much less the group, but a much smaller unit, namely, the gene. Thus, the struggle for survival is actually a competition among genes. Thus, in our example natural selection has favored the altruistic gene, not the altruistic wolf. This idea is also called the *theory of the selfish gene*, after British zoologist Richard Dawkins's 1976 book *The Selfish Gene*.

The theory of the selfish gene conceives of individuals—from amoebas to elephants, from *Boleteus edulis* to *Homo sapiens*—as machines constructed by genes purely as vehicles for their own survival through reproduction. We are all the survival machines of our own selfish genes. Although as a result of the struggle for survival it is not individual genes that perish or survive, but complete survival machines, nevertheless, selection operates at the level of the gene. The gene that finds a place for itself in a better survival machine is the gene that will survive. It is useful, of course, if the physical characteristics or behavioral strategies dictated by the gene contribute to the success of the survival machine, but for the gene, this is only of secondary importance. The gene has only a single aim: to secure its own survival. It is, of course, flagrantly teleological to put it this way. Perhaps it is better to say that *survival is the only reason for the gene's existence*, and so for the gene, survival is everything. There is enormous competition. Nature is constantly producing new genes by mutations, copying errors, gene splicing, and so on, but all the genes are engaged in "trying" to obtain a place in the best survival machines and to build survival machines that will particularly contribute to their own survival.

Darwin was ignorant of the molecular basis of genetics, although Gregor Mendel published the results of his experiments on peas with white and pink blossoms in Darwin's lifetime. The theory of the selfish gene, based entirely on the principle of natural selection, embodies Darwin's original ideas. Yet it not only explains the behavior of wolves, it also accounts for altruism, for a gene that demands self-sacrifice may, by saving other organisms housing copies of itself, promote the survival of its type of gene. (It doesn't "matter" to the gene which specific individuals survive. Two identical genes are equivalent.)

For the same reason, it may be useful for other genes in the survival machine to include this particular gene on their team. In this way their own chances of survival will also increase. The gene for altruism is interested only in the survival of itself and its likenesses; it does not care if the survival machine itself dies, along with one specific gene, itself, provided that its copies live on. Thus the selfish gene sacrifices "itself" to save "itself."

Competition Between the Two Theories

If nature is based on group selection, then it follows that nature oversees cooperation among the individuals of a given species wisely, even to the extent that an individual may occasionally sacrifice itself in the interest of the group. If the selfish gene theory is correct, then we are at the mercy of the interests of our genes.

Perhaps I should show my cards and reveal which theory—group selection or gene selection—I support. Over the past twenty years the advocates of the two theories have been engaged in vigorous debate. With such disagreement, how is a layperson to decide which of the two theories more closely embodies scientific truth?

The two theories suggest two radically different worldviews. According to the advocates of group selection theory, nature somehow takes care that individuals within the species (or at least groups within the species) usefully cooperate with each other, thus ensuring the survival of the species. It is as if a common goal—the good of the community—guides the actions of the individuals. According to the worldview arising from the selfish gene theory, the world is guided only by the immediate interests of individual genes, without any higher goal; the survival of a species is at best a mere byproduct, an incidental result, of genes pursuing their selfish ends.

Both theories are valid scientific theories in the sense that they can—at least theoretically—be tested by experiments, experiments that can confirm or refute the predictions of the theory.

Proponents of the two theories have conducted a large number of experiments. Looking over the last few volumes of the journals *Animal Behavior* and *Ethology*, I found about fifty papers that report experiments testing the theories' predictions. A variety of animals were used in these investigations, from sea horses to elephants. The results are highly diverse, which is due in part to the same types of difficulties as were encountered in the application of mixed strategies: A theory can give quantitative predictions only if the values of the possible outcomes are known. There are quite a few papers that appear to confirm one theory or the other (although the majority of the papers support the selfish gene theory), and in quite a few papers the experimental results do not confirm the predictions of either theory.

Physicists found themselves in a similar situation when they were attempting to determine whether the nature of light is corpuscular or wavelike. When experiments were carried out with particle detectors, light replied, "Yes indeed, I am composed of corpuscles; you can measure their impact." But when light was measured with instruments designed to detect waves, the experiments revealed beautiful interference patterns that only waves can produce. Now light was saying, "Yes indeed, I am made of waves." The reply depended on the question. As we shall see in Chapter 10, where we shall discuss this problem in greater detail, light actually has both characters. Or to be more exact, light is as it is, but it can act in both ways, depending on what we ask of it.

I am not sure that group selection theory and gene selection theory mutually exclude each other. If the question is asked whether natural selection acts at the level of genes or that of species, nature cannot help but follow light's example. It gives an answer to the question posed, regardless of what the real character of evolution is. Nature is as it is, but we are able to understand only as much as our human concepts permit. If the questions are formulated in an inadequate theoretical framework, then, as in the case of light, seemingly incompatible hypotheses may both be proven correct. Before the puzzle over light arose, no one had considered whether a physical object can be simultaneously waves and particles. Since the two theories logically excluded each other, physicists could not accept that in the case of light both theories are true. Such a conflict can only be resolved by a better theory, which for physics was that of quantum mechanics.

In many respects, group selection theory is similar to socialist theories of economics, while gene selection theory resembles free-market competition. This parallel will be discussed in detail in Chapter 9. In most countries the elements of both systems are present, with private enterprises based more or less on pure competition and state enterprises that conform more closely to the socialist model. This symbiosis is justified by the logic of economics.

Abstracting the concept of evolution from natural selection and conceiving it as a *universal natural force*, it may turn out that evolution is determined by a mechanism that on the one hand exerts its influence through natural selection affecting whole groups and on the other hand also acts at the level of the gene. It is also possible that these two types of effect are disjoined because of our limited conceptualization, while evolution simply is as it is, and, like light, only appears to be simultaneously of two natures.

Thus I declare for neither theory of evolution. If I must choose, I would say that the selfish gene theory is closer to my views, but not because I find the supporting evidence more convincing. The selfish gene theory seems at the moment to be a more useful tool for studying evolution, because the scientific puzzles that fall under its purview appear to be less hopeless. It seems that researchers within this theory are more likely to arrive at valid and interesting conclusions, which is why the majority of biologists are in this camp.

The two theories express two distinct worldviews, or paradigms, that apparently cannot be reconciled. Therefore, until somebody succeeds in integrating the two theories in a uniform (and certainly radically new) conceptual framework or it is demonstrated that one of the theories is invalid, researchers of evolution will be forced to commit themselves to one of the two theories if they want to do research. The way science is done forces their hands, even if they believe that neither theory is likely to be the last word on the subject.

The Struggle Between Hawks and Doves

The fundamental difference between the two theories of evolution is illuminated by John Maynard Smith's hypothetical example. Maynard Smith was the first to apply the methods of game theory in evolution research. This is how Dawkins introduces the idea through which game theory entered the biological sciences:

> Suppose that there are only two sorts of fighting strategy in a population of a particular species, named *hawk* and *dove*. (The names refer to conventional human usage and have no connection with the habits of the birds from whom the names are derived: doves are in fact rather aggressive birds.) Any individual of our hypothetical population is classified as a hawk or a dove. Hawks always fight as hard and as unrestrainedly as they can, retreating only when seriously

injured. Doves merely threaten in a dignified conventional way, never hurting anybody. If a hawk fights a dove the dove quickly runs away, and so does not get hurt. If a hawk fights a hawk they go on until one of them is seriously injured or dead. If a dove meets a dove, nobody gets hurt; they go on posturing at each other for a long time until one of them tires or decides not to bother any more, and therefore backs down. (*The Selfish Gene*, p. 75)

At the start of a hostile encounter neither party knows what strategy a selfish gene ascribes to its opponent. Let's assume that victory scores 50 points, injury results in a loss of 100 points, and in the case of an altercation between doves, the time spent on posing represents a loss of 10 points. Although we know from Chapter 1 that in this case doves would probably apply a mixed strategy, paying the average value for the disputed goods, let us now follow Dawkins's arithmetical scheme. The specific numbers are rather incidental; a similar analysis holds within a wide range of scores.

Let us prepare the game table, starting with the numbers above. If two hawks fight, then one gains 50 points, while the other loses 100. Let us assume that every hawk wins half of the fights and loses the other half. Thus the expected gain of each hawk will be the average of +50 and –100, namely, –25. If two doves engage in posturing, they will both lose 10 points for the time spent, while the winner will win 50 points, for a net gain of 40 points. The expected gain of a dove is thus the average of –10 and +40, that is, +15. If a dove and a hawk meet, then there is no fight. The dove scores 0, the victorious hawk +50. Thus, the table of the game is as follows:

| | | the other party | |
		hawk	dove
the one party	hawk	**–25**, –25	**50**, 0
	dove	**0**, 50	**15**, 15

We see from this table that this is a chicken type of game. Game theory (or rather its generalization by Nash) predicts that two points of equilibrium arise, namely, in the case when one party is a dove and the other a hawk. This, however, is meaningless in the present case, since if there are both doves and hawks in a population, then it cannot be avoided that sometimes hawks meet hawks and doves meet doves. It may become clear from this that von Neumann's theory of games cannot be applied directly here, even if we have a chicken-type table. It was John Maynard Smith who discovered that the ideas of game theory nevertheless can be applied to such a situation.

Rationality in the Selfish Gene Theory

We see, then, that the principle of rationality cannot be applied directly in biological investigations. Nevertheless, the logic of the situation is very similar to that which led to the discovery of mixed strategies in game theory. If a few doves appear in a purely hawkish population, the doves will fare well. They will always flee from the hawks, thereby breaking even, but when they meet one of their own kind, each dove will score 15 points on average. This is a much better deal than what the hawks get, who will lose an average of 25 points in every fight. Thus, the doves will begin to be fruitful and multiply. If, on the other hand, the population consists primarily of doves, the occasional hawk will be sitting pretty: In most of its encounters it will win 50 points without a fight. Thus, in a predominately dove population, hawk genes will rapidly proliferate.

This line of reasoning follows the logic of selfish gene theory. The dilemma that arises is somewhat similar to the one we pondered when we were discussing games—the loop of "I think that you think that I think that..." will never end. The idea of mixed strategies resolved the problem for games, and it will rescue the doves and hawks as well.

Let's see what happens if the proportion of birds in a large population is 7 hawks for every 5 doves. In this case, a hawk will meet a dove in 5/12 of its close encounters—winning 50 points each time— and it will meet a fellow hawk in 7/12 of the cases—losing 25 points on average. In sum, it will gain

$$5/12 \times 50 - 7/12 \times 25 = +6.25 \text{ points on average.}$$

Similarly, a dove in the same population will gain

$$5/12 \times 15 - 7/12 \times 0 = +6.25 \text{ points on average.}$$

The ratio of seven hawks for every five doves has led to a harmony similar to that achieved by the 1/9 proportion of bluffs in Chapter 5. Pursuing the matter further, if the proportion of hawks increases, the gain of the hawks will decrease, while that of the doves will increase; consequently, the doves will proliferate at the expense of the hawks until the ratio of hawks to doves returns to 7:5.

Thus, once the 7:5 ratio develops, it can remain stable for a very long time. If the selfish hawk and dove genes fight each other for survival, this is the equilibrium that will arise, and the proportion will vary only a little in either direction, since large deviations will be punished by the logic of the game. John Maynard Smith called a strategy leading to such stability an *evolutionarily stable strategy*.

Evolutionarily stable strategies are usually mixed strategies. The evolutionarily stable strategy in our above example can be brought about in nature in either of two ways: 7/12 of the population may always play the strategy of the hawks while 5/12 of the population always play that of the doves, or each individual of the population may act as a hawk or as a dove with the above probabilities—equilibrium will remain stable in either case. In the first case, the genes are simply distributed by nature in this proportion through natural selection. In the second case, natural selection favors those individuals whose genes prescribe the above probabilities of choosing one strategy or the other.

According to the selfish gene theory, evolution realizes a general principle of nature. This principle is that of *stability between competing individual genes*. Stability arises through evolutionarily stable strategies that are a particular variety of mixed strategy, in which the role played by the principle of rationality in game theory is taken over by the principle of evolutionary stability of gene populations. This amounts to a *higher-order rationality of the guiding principles of nature*. Like the principle of rationality, the principle of stability can also be realized only by mixed strategies. Furthermore, selfish gene theory provides an answer to the question of how evolution can create species harboring diverse and mutually competing genes that nonetheless survive stably.

Rationality in Group Selection Theory

According to the group selection theory, natural selection does not exert its effect on genes, but affects a population in which hawk genes and dove genes can both be present. According to this theory, the result of natural selection is to maximize the survivability of the entire population. In this case, struggle within a group does not make much sense, since it represents a loss for the group as a whole. It would be better to select the winner by drawing lots. In such a case, however, a population that actually engaged in animal fisticuffs would eventually wipe out the lot-casting population, since the fighting group has been selected for its fittest individuals. Ruthless selection operates *between* groups even if evolution sometimes creates the most peaceful solution *within* groups. The necessity of competition within a group is not ruled out by group selection theory, and thus in studying the struggle between hawks and doves, we can start from the very same hawk-and-dove table as in the previous thought experiment.

If there were no hawks at all in such a population, doves would live happily ever after, waxing rich with an average gain of 15 points per encounter, even at the expense of wasting some of their precious time on posing. According to group selection theory, it would appear at first that no hawks could exist in such a population, because natural selection would somehow eliminate them in the interest of the group as a whole.

If group selection theory really gave such a prediction, then we could scarcely consider it a viable alternative to gene selection theory, since it could not explain diversity within a species. But the situation is more complex. A world composed exclusively of doves cannot be considered the best of all possible worlds. If one-sixth of the group is hawks, we can calculate that each member of the group wins 16.6 points on average per encounter (this is the optimal proportion). This optimal proportion is called the *group selection optimum*. The optimal strategy of group selection theory can also be a kind of mixed strategy.

The proponents of gene selection theory believe that evolution can hardly bring about a group selection optimum, since such an optimum is not stable. The hawks fare much better than the doves. The danger of internal treason is always present: Doves might suddenly begin to behave like hawks, destroying the common optimum of the group. The proponents of group selection theory reply that once natural selection has affected the whole group, the individuals within the group have no alternative, and evolution is itself the force that creates the optimal proportions within the group. We have to investigate how evolution accomplishes this. Why, for example, do worker bees not rebel against their exclusion from the joys and responsibilities of reproduction?

Complex Strategies

The previous example of hawks and doves was intentionally simple. In nature, genes may determine complex strategies for their survival machines. It is possible that due to a mutation, a new gene appears in our hypothetical population of hawks and doves that prescribes the following strategy: Act like a dove at the beginning of each fight and assume that your opponent will do likewise. That is, begin posing, but if you see that your opponent is about to play the hawk and attack you, fight back! This gene could be called the *avenger*. If the population includes only doves and avengers, the carriers of the avenger gene cannot be distinguished from doves, because no fight ever occurs. If, however, there are also hawks in the population, the carriers of all three genes can be identified on the basis of their behavior.

Maynard Smith examined further strategies, like the *swashbuckler* and the *explorer*. The swashbuckler begins the fight as a hawk, but if its opponent fights back, he immediately runs away. The explorer behaves for the most part like an avenger, but it occasionally initiates a fight on the chance that the opponent is a dove. Using computer simulations, Maynard Smith studied how the proportion of these five genes changes in different populations through many generations. Depending on the original proportion of the individual genes, he found evolutionarily stable strategies, but also those for which no equilibrium arose but where the population remained in a stable state of oscillation. This is the case when the original population consists mainly of avengers and explorers. In this case, doves can always survive, although only in a small proportion, for if there are many doves, explorers will get the better of them, and avengers will come off slightly (but only very slightly) worse. The doves, hard hit by the explorers, will see their numbers decrease. However, when there are only a few doves left, the number of avengers will increase again at the expense of the explorers, since they are better off than explorers if there are only a few doves. This, however, makes the world a better place for doves. This oscillation may be sustained indefinitely. No stable equilibrium develops, although the appearance of a new gene may alter the situation.

The selfish gene theory is not supported by a pure mathematical theorem the way the principle of rationality is supported by von Neumann's theorem. Nothing guarantees the existence of a stable equilibrium. Nevertheless, an equilibrium does develop in most cases. The selfish gene theory also explains why in addition to stability, oscillatory behavior may sometimes develop.

In the case of group selection theory, the same value for the group optimum may often arise for different proportions of genes. Although group selection theory does not predict stable oscillation in these cases, it does explain why spontaneous changes of gene proportions sometimes occur within a population.

Genes may prescribe even more complex strategies for their survival machines. Not only may they prescribe what to do in a given encounter, but they may also determine long-term behavior. For example, they may prescribe that the individual apply the tit-for-tat strategy or any of the programs appearing in Axelrod's competitions of Chapter 3. It is also possible that a gene can differentiate between the prisoner's dilemma and chicken-type games and may prescribe alternative strategies for the two cases. Evolution may create genes that realize increasingly refined strategies, while it ensures the survival of the truly successful genes.

Group selection theory and gene selection theory explain diversity within species equally well. According to both theories, the higher-order rationality of evolution can be realized only by mixed strategies. The essence of the debate between the two theories lies in the nature of this higher-order rationality.

Applied to a hypothetical species consisting only of hawks and doves, the quantitative predictions of the two theories differ significantly as to the proportion of hawk and dove genes to be developed by evolution in the population. One predicts a 7:5 preponderance of hawks, the other a proportion of 5:1 in favor of the doves. We should be able to demonstrate this large difference experimentally, even if the numbers in our tables were well wide of the mark. However, the problem is that the proponents of the selfish gene theory carry out their experiments on totally different species from those of group selection theory. In the meanwhile, nature sighs and answers the questions limited by human understanding, acting like the rabbi in the joke who is adjudicating a dispute. After hearing the plaintiff's side of the story, the rabbi says to him, "I agree with you. You're right." But then he hears the defendant, and he says to him, "I agree with you. You're also right." Then the rabbi's wife, who has quietly been listening to the proceedings, interjects, "But Moishe, they can't both be right!" The rabbi reflects on this and finally replies, "My love, I agree. You're right, too."

9

Socialism and
Free Enterprise

While they are arguing over the spoils, you can snatch them.

According to an old joke that has made the rounds of Eastern Europe, the basis of capitalism is the exploitation of man by man, while with socialism the situation is just the opposite.

We shall study socialism and free-market competition not as political systems but as different economic principles. As we have indicated, we hope to show that if we study these two principles in their pure forms, their logic will correspond to the rationality of the group selection and gene selection theories rather precisely.

At present there is vigorous debate over which theory correctly describes how evolution operates. No intermediate solution has been forthcoming within the frameworks of these theories, and in the previous chapter we considered that the two theories might indeed be valid simultaneously. We can observe in economics how analogous principles can coexist.

There was a time when the virtues of free-market capitalism and socialism were fiercely debated. Today, however, neither capitalism based on free competition nor socialism in the sense of Marx can be said to exist, at least in their pure forms. Both principles have been tried and been found unfit. Systems that have proved to be more or less fit have arisen from combinations of these two principles. Economists call them *mixed economies*.

History has provided large-scale experiments with pure capitalism and pure socialism, often ending tragically, which happened to prove that these systems are unsuccessful in the long run. These experiments were not conducted by economists, but by societies, but economists had the opportunity to observe the experiments and even to influence them slightly. Biologists cannot make such observations, since evolution as a natural force is as it is, and it is not in the power even of politicians to test what the world would be like if evolution worked differently. Therefore, the large-scale political experiments in which different economic systems have been tried can be of use to biologists who study evolution.

On the other hand, biologists can conduct experiments that are unavailable to economists. They can observe organisms under artificial circumstances—just as Milinski did with his sticklebacks. Economists are precluded from performing such experiments because they would be ethically unacceptable.

All branches of science are sometimes faced with methodological gaps. In such cases it is usually preferable to close the gap by analogy with a similar scientific discipline than to enter into random speculation. Such an analogy must, of course, make sense, and in our case we must ask whether a force analogous to evolution is manifested in economic systems.

Economics and Evolution

If natural selection—the struggle for survival—is the principal mechanism of evolution, it should not be difficult to find analogies in economics. In nature, participants in evolution's Olympics compete for natural resources, while in the economic sphere, the battle is more literally for the gold. This difference might be fatal to our analogy, since, for example, customers sought by competing firms are capable of making conscious, reasoned decisions, while natural resources cannot. This difference will be accounted for in our models, but as we shall see, it does not affect their logical structures significantly.

In the case of participants in biological evolution, there is an obvious one-dimensional measure that indicates the fitness of an individual (or gene, or group): the number of offspring who survive to reproduce. This measure is computed differently according to the particular theory of evolution. Thus the selfish gene theory prefers to measure not the number of offspring, but the number of genes that are passed to the next generation. Nevertheless, each theory justly assumes that selection has a well-defined measure at its disposal, a measure that it "wants" to maximize. In economics, however, there is no such clear measure. Although the basic principle is to maximize profit, many other factors may intervene, from social sensitivity to the impudence of polluters. However, such secondary factors may also be governed by evolution, if we consider, for instance, the possible mechanisms of group selection.

There are factors in an economy that are not in the interest of any one participant but are nevertheless essential for the society at large, such as schools, roads, hospitals, insuring domestic tranquillity, providing for the common defense—just to mention a few. The necessity for lighthouses is a brilliant example given by Samuelson and Nordhaus. Lighthouses save lives and cargoes, but a lighthouse keeper cannot charge directly for services rendered, for how could he determine who has made use of his beacon? Furthermore, the cost of operating a lighthouse is independent of the number of customers. A beam that warns a hundred ships is no brighter than one that warns a solitary craft.

No economy can function without such communal enterprises, and natural selection cannot evolve them. Yet the same problem arose in biological research into the evolution of bee colonies. If evolution can develop division of labor such as that found in bee colonies, perhaps evolution could bring about economies that create communal values. These problems are analogous, so what might account for such different solutions? The existence of common services in an economy does not exclude the influence of the mechanisms of evolution. At the very least, these mechanisms resemble the principles of group selection.

The rules under which an economy operates are created by people, acting with free will. This may be an important factor mitigating against the existence of mechanisms of evolution in economics. Biological organisms have no role in the development of the rules under which they must operate. It was Darwin's ingenious idea that the development of species can be explained purely by the mechanism of natural selection. Consequently, concepts like "will" and "rationality" can be banished from evolutionary theory (though we shall see at the end of Chapter 10 that these concepts perhaps should not be banished completely, even if the theory of evolution is correct without them). However, these concepts certainly cannot be banished from the rules of economics.

Nevertheless, the goalless march of evolution seems to have created a purposeful human intelligence that creates economic regulations. If a person participates in an economy as if with perfect rationality, whether through reasoning, intuition, or just dumb luck, such behavior would in any case create the *appearance* of perfectly rational thinking, just as in nature the individuals and species that survive and flourish have behaved as if they had made rational decisions. Some economists, Milton Friedman for example, use this very "as if" phenomenon to explain why in their theories they consider people to be rational beings despite their all too frequent stupidity. Thus, just as undirected evolution has produced the appearance of rationality, so have the purposeful economic rules created by man.

The Invisible Hand

Perhaps the most influential work in economics is *An Inquiry into the Nature and Causes of the Wealth of Nations*, published in 1776, in which Adam Smith describes the principle of the invisible hand, according to which every individual, while working exclusively for his own personal gain, seems to be guided by a beneficent invisible hand to render the best possible service to the commonweal. Here are some frequently cited extracts from Adam Smith's book:

> It is not from the benevolence of the butcher, the brewer, or the baker, that we expect our dinner, but from the regard of their own self-interest. We address ourselves, not to their humanity but to their self-love, and never talk to them of our own necessities but of their advantages. (p. 14)

> He generally, indeed, neither intends to promote the public interest, nor knows how much he is promoting it.... He intends only his own gain, and he is in this, and in many other cases, led by an invisible hand to promote an end which was no part of his intention.... By pursuing his own interest he frequently promotes that of the society more effectually than when he really intends to promote it. (p. 423)

By explaining its rational nature, Adam Smith became a prophet of free competition. Economies certainly existed before the development of the free market, and even then, high levels of equilibrium developed, as in the so-called Asian mode of production, which survived almost unaltered for thousands of years. These systems, too, could be the product of evolution, perhaps of a form in which emphasis on group selection was greater. When free competition developed, however, this represented perhaps the appearance of a new "gene" in the economy, shifting evolution's emphasis to that of gene selection.

Adam Smith never exactly verified his invisible hand principle. Although we saw in Chapter 7 that competition may promote cooperation, until John von Neumann appeared, nobody knew how Adam Smith's very effective theory could be verified, even in part, by scientific methods. Adam Smith's theory was purely intuitive. The theory proves to be correct only under conditions of perfect market competition, and Smith gave examples from ancient and contemporary history of how well-intentioned government intervention could produce an adverse effect.

Von Neumann's theorem and its generalizations to multiperson games in economics have largely supported Adam Smith's intuitive insights. The principle of rationality, according to which every player seeks to promote his own pure interest and assumes the same of his opponents, may engender prolonged and stable equilibrium.

Yet it was game theory itself that indicated the limitations of Adam Smith's theory. In cases of non-zero-sum games, the invisible hand can lead far afield from the common optimum. For instance, in the prisoner's dilemma (and its multiperson version, the problem of the common pasture), the invisible hand leads to mutual defection—thus to catastrophe—and this is the only Nash equilibrium point. In such situations the government may be required somehow to herd the participants in the economy into mutual cooperation.

Not only was Adam Smith's book a forerunner of important trends in economic theory, it also presaged Darwin's theory of evolution, which was developed almost a century later. According to Darwin, the basis of change in nature is the struggle for survival of completely selfish individuals, and it is this struggle that has led to biological diversity on earth, to the development of species. Evolution is the invisible hand that governs this whole process, and its method is natural selection. The theory of Adam Smith, however, applies only to that part of evolution that is represented by the theory of gene selection. If group selection mechanisms, too, are involved in the operation of biological evolution, then these must appear somehow in the economy as well, even in economies of purely free competition.

Theories of Equilibrium

Among the players of the game "economy" are manufacturers and consumers. Every manufacturer has his own set of conditions that determine what he can produce at what cost. Thus the possible pure strategies—the goods that he can produce—are available to every manufacturer. Most manufacturers do not apply their entire capacity to produce a single product: They play a mixed strategy. The outcome a manufacturer can achieve by playing a particular mixed strategy depends on his costs of production and on the price the product can fetch in the marketplace. If the cost and price of every product are known, then the outcome for any mixed strategy can be calculated exactly.

The players taking the role of customer also have their agendas. They know how much money they have, and they also know the degree of satisfaction they can achieve by buying a particular commodity at a particular price. Some products can be substituted by other products. Moreover, customers do not spend all their money on a single product: They, too, play a mixed strategy.

Naturally, manufacturers are also customers, for they buy raw materials and services, but let us not complicate things too much. Let us consider each manufacturer to be two players: one manufacturer and one customer.

If all products were available in the marketplace in unlimited quantities and at a fixed price, then we would be able to construct the game table and study it by the usual methods of game theory. In economics, however, the situation is more complex. Customers can buy only as much as manufacturers produce, and prices, far from being fixed, depend on demand and supply. Therefore, the table of the game has to be modified accordingly. If the strategies of all of the players are known, we can tell which products will give rise to a surplus or a shortage. But before we can calculate this, the market will have reacted in two ways. On the one hand, the price of a product in short supply will rise, while the price of surplus goods will fall. On the other hand, as prices change, manufacturers and customers will change their mixed strategies. Manufacturers will attempt to produce more of the higher-priced goods, while customers will find alternatives to purchasing expensive products.

We are caught in the same infinite loop in which we pondered, "I think that you think that I think that...." Can market equilibrium be established? If so, under what conditions?

The Nobel laureate economists Kenneth Arrow and Gerald Debreu found the answer to this question, which is essentially the generalization of von Neumann's theorem to the above game. According to the Arrow–Debreu theorem, there exists a Nash equilibrium of the above game under quite general conditions. (Economists prefer the term *weak Pareto optimum* for essentially the same concept.) An equilibrium can develop in which no player can increase his profit unilaterally simply by changing his strategy. The Arrow–Debreu theorem, which has become an important method of analysis in modern economic theory, is also called the *general theory of equilibrium*.

At first sight, this theorem completely vindicates Adam Smith's views. The invisible hand triumphantly creates a stable economic equilibrium acceptable to all, and this is guaranteed by the implacable logic of mathematics. The devil, however, as the saying goes, is in the details. I said above, "under quite general conditions." Although these conditions are really quite general, they are not fully met in any real economy. For example, the following conditions are necessary for the general mathematical validity of the Arrow–Debreu theorem: There can be no effects within the economy whose origins are outside the economy; economic activity has no effect beyond the economy; prices and wages are completely flexible; there are no monopolies. In addition, quite a few further technical conditions are necessary, such as the "law of decreasing profit." We will not discuss the details of these conditions here.

If a government wants to create the purest possible free competition, that is, if it wants the greatest possible invisible hand effect, then these conditions must be satisfied, even if by significant restrictions on free competition. Samuelson and Nordhaus conclude:

> When the checks and balances of Darwinian perfect competition are absent, when economic activity spills over outside of markets, when incomes are distributed in politically unacceptable ways, when people's demands do not reflect their needs—when any of these conditions arises, then the economy is not led by an *invisible hand* to an optimum position. Further, when a breakdown occurs, the carefully designed and restrained intervention of government may improve economic performance on this imperfect and interdependent globe.

In the above quotation Samuelson and Nordhaus use the expression "Darwinian perfect competition." They do not consider the possibility of group selection, although as we have seen, Darwin himself did take it into account. The constraints and compensating factors of "Darwinian perfect competition" do exist in nature, namely, group selection mechanisms, although we do not know exactly what they are. Whatever they are, perhaps their imperfect, earthly counterparts in human societies are governments. Once a government is in power, however, it can interfere with economic processes for a variety of reasons.

Planned Economies

An economy can be guided by more than an invisible hand "supplemented" by government intervention. If this were the only role a government played, it would be possible to consider it as a slightly visible part of the invisible hand, mildly influencing the economy's activity or perhaps improving its efficiency. In the former socialist countries of Eastern Europe, National Central Planning Boards and the National Offices of Supply and Price-Fixing were all-too-visible hands that fundamentally determined economic processes. It is characteristic of purely socialist economies that the government determines the allocation and use of resources and orders the participants in the economy to follow the state's economic dictates.

One may be inclined to look at such command economies as irrational utopias or, in Eastern Europe, as a nightmare recently ended. Yet such economic systems are also the result of rationalism, indeed, rationalism carried to its logical extreme. The system of planned economy is based on an a priori rejection of market equilibrium. A planned economy attempts to optimize the profit of the community by government fiat. Paradoxically, it employs the same mathematical methods as those used to prove the invisible hand's equilibrium. This is not very surprising if we consider that group selection theory and the selfish gene theory both lead to applications of game theory and mixed strategies.

We have seen that the Nash equilibrium is far from being optimal for the whole community—often such an equilibrium doesn't develop at all. Let us imagine that an economy has twenty participants whose economic activity consists in playing a one-shot prisoner's dilemma game with each of the other players according to the table below. Naturally, as an economic model it is completely absurd, but it will serve to show the essence of our argument.

| | | player 2 | |
		cooperates	competes
player 1	cooperates	**3**, 3	**0**, 10
	competes	**10**, 0	**1**, 1

This table resembles the game studied by Axelrod, but here the temptation to compete is even greater, for now the competitive player who meets a cooperating sucker will win 10 units instead of only 5. In a game where both players cooperate, the two players gain a total of 6 units. If both of them compete, together they will win 2 units. If one of the players cooperates and the other competes, the two players will together gain 10 units. Thus, the total production of the economy will be increased by 6, 2, and 10 units, respectively.

Each player plays with 19 other players. Thus 190 games are played altogether. If all of the players cooperate all the time, the players will gain

$$190 \times 6 = 1140 \text{ units.}$$

That is what application of the golden rule leads to. According to the principle of rationality, the Nash equilibrium will be reached only if everybody competes. In this case, however, the players will gain only

$$190 \times 2 = 380 \text{ units.}$$

This is predicted by gene selection theory, and Adam Smith's invisible hand also leads the players to this result. But what happens if 14 players cooperate all the time, while the remaining 6 players always compete. In this case, in 91 of the 190 games both players will cooperate, in 15 games both players will compete, and in 66 cases one of the players will cooperate and the other will compete. Thus, altogether the players will gain

$$91 \times 6 + 15 \times 2 + 66 \times 10 = 1236 \text{ units.}$$

In this prisoner's dilemma, where the temptation to compete is particularly great, the common optimum is not reached by everybody cooperating. The situation is closer to that found in games of chicken: Here again the categorical imperative, or group selection theory, prescribes a mixed strategy.

What can an economist do who has unlimited power in a command economy and who always works for the common good? The best he can do in the interest of the common good is to prescribe a competitive strategy to 6 of the 20 players, while compelling the others to cooperate. In this way the gross national product will be almost ten percent higher than it would be with everybody cooperating, not to mention the case where everybody competes.

The realization of such an optimum in an economy runs into several theoretical and practical roadblocks. One obstacle is that such a system could hardly be maintained for long in the real world, unless the government were to take over the distribution of goods; that is, it would require perfect communism. But even then, it would be very difficult to prevent the economic units that are consigned to competition from exchanging their advantageous positions for cash.

It is both a theoretical and a practical obstacle that often it is difficult to determine just what needs to be optimized. There was a period in Soviet history when the dishes manufactured by the state-owned porcelain factories began to grow thicker and heavier. Why did this happen? Just as they did in the iron and steel industries, state planners had set porcelain manufacturing goals in tons of output.

The practical obstacle is that our omnipotent economist needs to collect and analyze enormous quantities of information—that is, he must be omniscient as well. Not even today's computers could carry out such an analysis. We're talking about equations with billions of unknowns, and this would take years even with our fastest computers. But even if it could be analyzed, to collect the necessary data would be impossible, in large part because the interests of most participants in the economy are to keep information secret. This was always the case in the Soviet Union, but even in the United States the government was able to collect only a small fraction of the data it sought for a 1974–75 large-scale energy model for the country.

Experience shows that it is impossible to complete such programs effectively. Furthermore, it is not the only job of government to regulate the market (that is, to fulfill the conditions of the Arrow–Debreu theorem). Government should also promote the common interest—undertake those projects that the majority considers necessary or desirable, from decreasing pollution to the maintenance of lighthouses. It is for this reason that mixed economies have developed.

The Diversity of Mixed Economies

Mixed economies are governed partly by the participants in the economy, partly by government-directed public institutions. One of the components—the totality of the participants in the economy—directs the functioning of the economy by the market's invisible hand. Its operation is described quite precisely by the theory of gene selection. The other component—government—directs the economy in two ways. On the one hand, it can subtly influence the operation of the invisible hand through market regulation and financial incentives, while on the other hand, it can direct the conduct of public affairs in such a way as to provide a positive influence on the marketplace. The government, then, is the component of the economy to which group selection theory may be applied.

The relative weights of the two components are quite different in different countries. In the United States gene selection predominates, while in Sweden, for example, group selection has primacy. In every democratic country free elections determine in part the relative weight to be given to each component. Nevertheless, in long-established democratic countries radical change rarely occurs. The Swedish political right would be considered left-wing in the United States. In these countries the proportion of the two components fluctuates within rather narrow limits, although quite large differences can be found among countries.

The amount staked in the fight between two male elephant seals is usually very high: The winner will possess a whole harem. No wonder the fight is generally violent, and the loser is usually severely injured. In this case, the selfish gene theory seems to be operational. Yet in the case of ants, group selection theory seems more apt. Sticklebacks are somewhere in the middle. The genes of individual organisms also determine to some extent the degree to which the two principles of evolution are present. The situation is similar in economics: The attitudes, culture, national character, education, and constitution of the labor force determine the proportion of the two components of a mixed economy.

The biological parallel is not entirely without foundation. The American biologists Charles Lumsden and E.O. Wilson noticed that our social milieu is composed of elements that can be substituted for one another, just as our biological self is constructed from the blueprint of our genes. For instance, we can choose the clothes we wear, the stories we tell our children, and the strategies we apply in solving problems. It is as if we selected our own genes for skin color, height,

and physique. Although social attributes are not inherited genetically, nonetheless they are passed on as a kind of cultural heritage,
and they can thought of as if they fight their own struggle for survival. Lumsden and Wilson called these elements *culture genes* and
quite successfully applied the mathematical methods developed for
genetics to them. The equations of evolutionary biology applied to
culture genes were able to model a number of cultural characteristics, from the evanescent whims of fashion to our most enduring cultural monuments.

Significant differences are to be found between the cultural gene
pools of different countries. As a result, evolution can bring about
different types of mixed economies. The general laws of economics,
however, hold generally. In this there are no real differences among
countries.

Likewise, we have good theories about the mechanisms of gene selection, and they can be applied to any species. Different species,
however, engage in very different forms of combat, although the aim
of every struggle is the same: survival. The individuals of one species
may fight to the death to settle a dispute that is resolved in another
by nonviolent posing. The behavior is determined by the genetic
makeup of the species. The general conditions under which the fight
takes place are determined not by gene selection, but rather by the
mechanisms of group selection—although we have much less of an
idea how this works.

According to our analogy, the counterpart of animal species is the
economies of countries. If a country wants to direct its economy to
influence the invisible hand (which is necessary, as we have seen),
this may require completely different steps in different countries, for
a method that proves effective in one country may not work at all in
another.

The Logic of Evolution

In analogy with economics, it may be that evolutionary theory does
not have to resolve once and for all the debate between the gene selection and group selection hypotheses. Perhaps the two theories
simply throw light on two *equally valid* aspects of the mechanism of

evolution. In economics, those systems appear to be favored in which both principles can function simultaneously.

It is as if evolution itself were playing a mixed strategy with gene selection and group selection, applying the various proportions of the two to both economies and species.

Nature can apply a mixed strategy in two ways. It can create populations of individuals each of whom plays one or another pure strategy, or it can create individuals who realize the appropriate mixture of strategies within themselves, behaving now one way, now another. Experimental evidence points in both directions. Biological evolution most likely uses the latter method to combine the two theories of selection: Both mechanisms affect every species, although their proportions may vary among species.

Among the mechanisms by which the natural force we call evolution exerts its influence, the first that we considered was natural selection. We have seen that natural selection may have many forms in nature, such as group selection and gene selection, and it is also possible that there are further, hitherto unknown, forms. Now, we may ask by what mechanisms evolution determines the proportions among the different forms of natural selection. This question is not ripe for systematic investigation in biology, but the example of economics may provide some hints.

In an economy, the proportions of the two types of regulating mechanisms can be subtly controlled by democratic systems through regular elections. These elections are ultimately about the economic direction in which a government should move. For all the theoretical problems with this assertion, problems such as the limited rationality of voters (to what extent do irrational forces operate beneath the apparently rational surface of democracy), for economies, democracy has shown itself nevertheless to be the most effective means of evolution thus far.

This does not mean that the higher-order mechanisms of evolution in nature are some abstract form of democracy. Most likely they are not. We can say only that democracy realizes some of the unknown mechanisms of evolution successfully, even if unknowingly and unintentionally. After all, democracy is a product of evolution itself. Perhaps it is nothing more than a highly successful cultural gene. Its success is due perhaps to its ability to guarantee social and economic stability, thereby promoting the survival of the selfish cultural gene. Perhaps it has proven effective *because of* its superficial rationality and deeply hidden irrationality. The reader may take this last remark with a grain or two of salt, though in the third part of the book we shall talk about such topics in discussing human thinking.

What ultimate principle governs evolution itself? Perhaps such a final principle is a higher-order, hitherto undiscovered, form of rationality that mixes the already discovered forms of rationality, namely, the principle of rationality, the principle of stability, the categorical imperative, and perhaps other principles as well. But perhaps this higher-order principle transcends rationality, a principle that exceeds human understanding.

10

Games Elementary Particles Play

Even an adolescent's moods are more predictable than the
motion of an electron.

In 1900, after completing his studies at the Federal Polytechnic
Institute in Zurich, Albert Einstein applied throughout Europe, without success, for an academic position. He then accepted a modestly
remunerated but not very stressful job at the patent office in Bern,
which provided a suitable environment for him to think about both
the direction of his life and the conundrums of physics. In 1905 he
published three papers, each of which solved one of the three funda-

mental problems of contemporary physics. Each of these papers was worthy of a Nobel Prize, and indeed, in 1921 Einstein was awarded the Nobel Prize, not for developing the theory of Brownian motion nor for his revolutionary special theory of relativity (which even then was still controversial), but for solving the problem of the photoelectric effect. We shall soon look into the essence of this problem.

When the theory of relativity finally came to be recognized as Einstein's great achievement of that year—despite or because of its incomprehensibility—the world laughed loud and long at the Nobel committee for not having recognized the theory's importance. However, looking back with the perspective of almost a century, it was after all perhaps the idea that won the prize that has proved to be the most fruitful.

The theory of relativity was the capstone of classical physics, whose system of ideas was now supplemented by an ingenious new principle. It was a revolutionary idea acceptable to the established order: exhilarating, clever, effective, but not one to turn the old worldview upside down. Newton's reassuring, deterministic worldview could be still maintained.

The solution of the problem of the photoelectric phenomenon, on the other hand, diverted the science of physics into a completely new channel—*quantum mechanics*. Einstein himself was never able to accept the worldview that this new science required. He remained an ingenious and highly respected critic of the new physics. Everybody paid attention to his opinions, and as a result of vigorous debate only the most rigorously formulated propositions of quantum mechanics were able to survive. In this way he contributed to the remarkably rapid development of modern physics.

Perhaps the most convincing argument in favor of the strange, counterintuitive world of quantum mechanics is that its predictions are verifiable in the world of very small particles, just as Newton's mechanics operates in the macroscopic world. The success of the atom bomb is just one, though perhaps not very fortunate, example. But microelectronics, laser techniques, and many other technical achievements could not have been developed without quantum theory. The American physicist Leon Lederman, who shared the 1988 Nobel Prize in physics for his work on neutrinos, once estimated that over twenty-five percent of the gross national product of the advanced industrial nations comes from products based on a knowledge of quantum physics.

Although we quickly put into practice the technical achievements made possible by quantum physics, it has only slowly penetrated our worldview. We are reluctant to accept that the universe may really be the strange place that quantum physics describes.

Einstein's main counterargument, which he maintained until his death, was, as he put it once in a letter, "God does not play dice." And on account of that article of faith he continually attempted to squeeze the genie back into the bottle from which he himself had released it. He attempted to develop alternative theories from which it would follow that elementary particles, and thus the entire universe, are after all not governed by blind chance. The course of physics, however, has not been kind to Einstein's point of view.

Using the methods of game theory, the basic idea of quantum mechanics can also be phrased as follows: *The elementary particles realize the idea of a mixed strategy*. Perhaps Einstein would have found the following formulation more acceptable: "God so loved the world that he supplied it, from elementary particles to human consciousness, with mixed strategies." As we know from game theory, mixed strategies often provide the only possibility for reconciling diversity and stability.

A number of excellent popular books on quantum physics have been published. The review of the subject in the next few pages has no desire to compete with these. I will utter not a word about a number of high points of quantum physics, including my personal favorites—Heisenberg's uncertainty principle and the quarks. My aim here is to cast light on the intimate relationship between quantum physics and game theory.

The Dual Nature of Light

According to our intuition, light travels straight as an arrow. After two hundred years of systematic and thorough experimentation, however, physicists have proved that light behaves like a wave. You may think of the waves on the surface of a lake when you throw a pebble into it or the waves that travel along a plucked guitar string.

Compared to such waves, the wavelength of light is very short (the range of visible light is around a thousandth to a ten-thousandth of a millimeter), but it has the same properties as its longer-wavelength counterparts. Two light waves can reinforce or cancel each other, or produce interference patterns like those made by two pebbles dropped into a lake a short distance apart.

In a certain sense, waves of light are simpler than waves of water or waves on a guitar string. Waves of water do not travel at a uniform speed: Larger waves travel faster. Light travels at a uniform speed in a vacuum—about 3.0×10^8 meters (186,000 miles) per second—regardless of its wavelength.

The uniform speed of light caused no mathematical problems for physicists. Difficulties arose only in investigating the so-called *photoelectric effect*. The essence of the effect is that if strong monochromatic light is cast upon metal, the metal will emit electrons. Qualitatively, such a phenomenon is predicted by classical physics. Light waves transmit energy, and that energy might easily detach a few electrons. However, the equations of classical physics predicted totally different departure energies for these electrons from what was measured.

This contradiction was resolved in two steps. First, Max Planck hypothesized in a totally different investigation that perhaps light does not arise continuously as a wave, but in small doses, in so-called quanta. This assumption solved the problem that was bothering Planck at the time (understanding the radiation of so-called blackbodies, but this is not important for us now) but could not by itself explain the photoelectric effect. Einstein's idea was still necessary, namely, that not only do materials *radiate* light in quanta, but they also *absorb* light in quanta. Thus, the energy carried by the wave not only arises in quanta, it also arrives in quanta, despite the fact that between departure and arrival the energy seems to be carried by a continuous wave. This assumption may have seemed absurd, but it resulted in a simple formula that experimental results have confirmed exactly.

The contradiction to our intuition in Einstein's theory is that if light is a wave, then how can it set out and arrive as quanta. Existing in quanta is a characteristic of particles, not of waves. That light arises in quanta is a strange and difficult concept, although it can be imagined somehow by the analogy of the pebble thrown into the water. Arrival of light in quanta, however, totally contradicts how we imagine wave behavior. Yet in Einstein's formula wavelength was a factor, a factor uninterpretable in a particle theory of light.

Using an analogy of Douglas R. Hofstadter's, the situation is as follows: A frog leaps into a pond, and as a result, waves appear. The waves are propagated as they should be, but just before reaching the shore, they cease to be waves. The water suddenly becomes calm, the wave is transformed into a frog, the frog jumps out of the water onto the shore, and then the frog kicks some of the rocks lying there. The longer the wavelength, the smaller the frog that jumps out, and conversely: A very small, quickly vibrating water surface gives rise to huge frogs. The shape of the wave depends on the size, inertia, and movements of the frog as it leaps into the water, but the size of the frog jumping out of the water depends only on the wavelength and not on the individual characteristics of the frog. The key to the analogy is this: The frog is the light; the shore is the metal surface on which the light is cast; the rocks on the shore are the electrons in the metal. Oh, and one more thing: The wave that arose from the frog and became a frog again—that is also light.

Although the analogy is preposterous, the picture it paints conforms to experimental results (on light, not frogs) quite exactly. The picture will become a little more complicated later, when as the wave is being propagated, the frog is also present, and when the frog reaches the shore, the wave is also present in the frog jumping out of the water. But this is to be expected. How otherwise would the wave anticipate its arrival on shore? It must be constantly prepared to change into a frog.

If we have to, we'll somehow get accustomed to this absurdity of light's behavior—just as since Copernicus we got used to the absurdity that the earth revolves about the sun, or since Galileo that in a vacuum a heavy object drops no faster than a light one. Considering the analogy with the frog, however, it seems that perhaps we shouldn't consider light as a wave at all. The only evidence of waves is that there is a relationship between wavelength and the size of the frog. Yet it follows from Einstein's formula that the size of the frog depends *exclusively* on the wavelength. Perhaps we could replace wavelength by some other abstract concept that could be applied to particles as well. How strong is the evidence that any model of light must include wave characteristics?

Two-Slit Experiments

One of the most important experiments proving the wave characteristics of light was conducted in 1804 by Thomas Young, a medical doctor who on the side was interested in the nature of light (he also helped decipher the hieroglyphics on the Rosetta stone). He cut two closely spaced parallel slits in an opaque screen. He then projected monochromatic light (light consisting of a single wavelength) on the screen, allowing the light that passed through the slits to fall on another screen. The effects were visible to the naked eye. If Young covered one of the slits, an image of the other slit was visible on the second screen, with blurred edges due to the scattering effect of the slit's edges. If he covered the other slit, he obtained the image of the first. Now, if light were corpuscular in nature, then with both slits open we should see the images of the two slits on the second screen. Spectacularly, the result was not two images at all. A band of dark and light stripes appeared on the screen. The number and pattern of stripes depended on the distance between the two slits and on the color, hence the wavelength, of the light.

These typical interference phenomena are easy to detect in water and sound waves. The cause of the interference is that the two beams of light coming from the slits meet and form an interference pattern of reinforcement and cancellation. Since the points on the second screen are at different distances from the two slits, different portions of the interference pattern arrive at different points on the second screen. It can be calculated easily what stripes should appear where on the screen if light is really a wave. In the experiments, the stripes appeared exactly as predicted by the calculations. Thus, light behaved just as expected if it were a wave.

Theoretically, this experiment can be performed by electrons, too, although it requires more complex technical apparatus. However, it is worth spending time and money on conducting such an experiment only if we suspect that electrons may also have wave characteristics. We shall discuss the origin of this suspicion later, but for now

it is worth knowing the outcome of the experiment. To be histori-
cally precise, the exact experiment that I shall describe has never
been conducted. Davisson and Germer actually performed much
more complex experiments in the 1920s, from the results of which
the outcome of the more simple experiment follows.

Let us assemble the necessary equipment. We shall need an electron gun with which we can shoot electrons one by one in the direction of the screen. We also need an apparatus that can record the hits
made by the electrons. A Geiger counter, which measures radioactiv-
ity, will serve the purpose perfectly well.

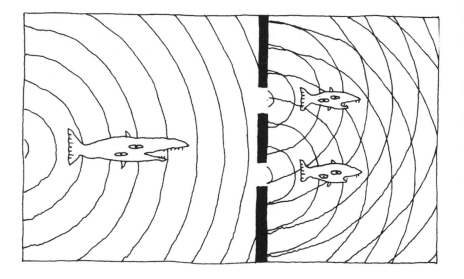

Two screens are used in the experiment. The first screen is made of
lead, because electrons cannot pass through lead. There are two slits
in this screen, quite close to each other. The second screen is
equipped with electron detectors, and in this way the locations of the
hits are recorded. First, one of the slits is covered by a lead plate,
then the other, and then both slits are left open. The electrons are
fired one by one by the electron gun, which swings rapidly from left
to right, thus, the two slits, and often the lead screen itself, are hit
one after the other. If only one slit is open, the image of one slit
gradually becomes distinct on the second screen; that is, there is a
single stripe with blurred edges. However, if both slits are open, the

image on the second screen is exactly the same as that received by Thomas Young: a row of alternating dark and light stripes!

The electrons were fired one by one, spaced so that an electron was fired only after the previous one had reached the second screen if it passed through one of the slits. Thus, the interference pattern cannot be explained by interference between electrons. Yet each electron caused only a single hit, thus in this respect behaving like an ordinary particle. *How could a particle know while passing through a slit whether the other slit was open?* Yet somehow it did "know"; otherwise, how could it have arrived at one location—in the image of the slit—when one slit was open, and at another location—among the dark and light stripes—when both slits were open?

The frog analogy may help us to understand this phenomenon. The firing of the electron gun represents the moment when the frog leaps into the water. Then the frog, or the electron, behaves as a wave, and a wave perceives whether one or two slits are open: It passes through only one or both accordingly. Each open slit is the source of another wave, just as when water waves, arriving at a narrow gate, begin to spread again from the gate as if they had originated from there. If only one slit is open, then only a single wave travels further to the second screen, while if two closely spaced slits are open, the waves originating at the two slits will interfere with each other. Just before arriving at the shore (the second screen), the wave itself disappears and is transformed again into a frog, that is, an electron particle, which is detected by the Geiger counter. Thus we obtain several interference stripes on the second screen if both slits are open. There is yet another mental leap we have to make—the very leap that shows how our present tale fits into the theme of this book. As it turns out, our frogs are actually probabilistic frogs that realize the mixed strategy of electrons.

If You Ask a Stupid Question . . .

While the frog analogy may help in understanding what may have caused this perplexing result, nevertheless, one may justly wonder where the electron had disappeared to while it was being a wave. What would the detectors have detected if they had been built into the first screen? We could carry out such an experiment, and we would see that the electrons fired by the electron gun sometimes hit one slit, sometimes the other slit, and most often the lead screen.

It is as if the electron passed wavelike over the first screen, and wherever it found a slit, passed through. If it found two slits, it passed through both. However, we cannot measure any waves—not at the slits, not on the first screen, not on the second screen—because as soon as we attempt to measure a wave, the wave immediately changes into a frog. This is why we can never tell where the electron *would have* hit the second screen if we had not looked at it on the first screen. We have no way of answering questions like, Where was the electron before hitting the screen? We cannot even tell which slit it passed through when both slits were open. The question itself is not well-posed, and nature (or the Creator) only shakes her head in disapproval, while murmuring between rolls of the dice, "If you ask a stupid question, you get a stupid answer."

No question about it, the behavior of the electron is weird. If we happen to catch it at one of the slits when the other is closed, it hits our detector like a cannonball and noisily declares, "I'm striking here. I'm nowhere else. Other slit? What other slit?" However, if we do not so catch it, it will mockingly thrust its tongue at the second screen, saying, "I was at the other slit, too. I saw that it was open. How on earth would I know otherwise that I have to arrive at one of the many interference stripes?" With such experimental results, it is no wonder we have to reconfigure our worldview radically if we are to understand the phenomena of the quantum universe.

We now have an instrument with which we can perform Young's experiment with individual photons (the quantum form of light). We emit the photons one by one, not in quantity as Young did. This way, the corpuscular nature of light becomes visible in addition to its wave character—as predicted by Einstein—and even the simultaneous presence of both can be shown. Today, this is no longer surprising: If the theory were proven fundamentally false, we could throw out all our computers and CD players. Likewise, Newton's physics was not proven false; rather, its limits became clear. However, the *worldview* derived from considering that Newtonian mechanics is a general law of nature was shattered. Nevertheless, we can still safely sit in an airplane designed on the basis of Newtonian physics.

Schrödinger's Equation

The year is 1925, and the Vienna-born physicist Erwin Schrödinger is attempting to rescue classical physics. He is studying electrons and has started from an idea similar to ours when we wanted to find out after learning about the frog analogy how strong the evidence really was in favor of light possessing wave characteristics. Perhaps we could find some abstract concept to replace that of wavelength that could be applied to particles as well, and in this way we could save the idea of light possessing a uniform character. Perhaps in reality light consists of particles, and its wave character is only a deceptive manifestation. The situation with the electron was just the opposite: Physicists had always thought that electrons were of a corpuscular nature. Schrödinger thought, conversely, that if he conceived of light as having only a wave character, it might turn out that its corpuscular nature is sheer illusion, and there are only matter waves, waves for which a particle-like abstract concept could perhaps be found.

With this goal before him, Schrödinger set up an equation to calculate the energy level of the electron in which electrons are waves, not particles that occasionally behave as waves. In this equation an electron is characterized by a function that includes every piece of measurable information about the electron. He denoted this function by the Greek letter ψ (psi) and called it the *wave function*. Schrödinger's equation gives the spatial and temporal changes of the wave function if the characteristics of the forces affecting the electron are given.

Schrödinger's derivation is mathematically flawed. Almost all of its steps are incorrect. Nevertheless, this derivation stands as one of the great feats of physical intuition. The derivation of the formula (I would rather say the intuitive method of its construction) uses almost all the important achievements of classical mechanics, supplemented by the then known relationships of quantum phenomena. Yet the formula is amazingly simple. It can be applied not only to electrons, but to any other particle, and even to quantum-mechanical systems consisting of several particles, and hence to atoms and even molecules. As Robert Oppenheimer said, it is a formula "perhaps one of the most perfect, most accurate, and most lovely man has discovered." If Schrödinger's equation is applied to a group of many particles, that is, to a macroscopic object, it will lead to Newtonian mechanics as a limiting case.

This characteristic is particularly lovely, since it shows that Newton's physics may still remain valid for macroscopic phenomena.

Schrödinger's equation also explains nicely how one part of an electron bundle can pass an obstacle (more exactly, a damping force field) while another part cannot. If the wave function is higher than the obstacle, then part of the substance represented by it can naturally pass, just as parts of the waves on a choppy lake can pass over a dam. The percentage of the material that can pass the barrier can easily be calculated by Schrödinger's equation. For instance, in the two-slit experiment, we can calculate what fraction of the electrons will be detected on the second screen.

Although Schrödinger explained in detail why he composed his equation as he did, the equation can hardly be considered a physical law that follows from the other laws of physics through a chain of logical deduction. Rather, it can be thought of as an axiom, a physical hypothesis, a statement requiring no proof, because that is how the world operates, no matter why it does so, just as Euclid's axioms describe the behavior of points and lines in a plane. A famous Hungarian professor once wrote Schrödinger's equation on the blackboard at the start of the first lecture in a course on quantum mechanics. Then he turned to the class and said:

> Ladies and Gentlemen! This is the famous and celebrated Schrödinger's equation. I know that nobody understands this equation. You don't. I don't. Nor does Mr. Schrödinger. But do not let this bother you. I will write this equation on the blackboard at the start of every lecture, and I will explain how it can be used. And sooner or later you will get accustomed to it.

Probabilistic Frogs

Researchers in quantum physics soon got accustomed to Schrödinger's equation, and many even grew to like it, because it proved its usefulness repeatedly. For one brief, shining moment it really seemed to be true that everything in the universe was a wave, and that the existence of particles was only an illusion. In this developing brave new worldview, however, there were stubborn problems not willing to yield before the might of Schrödinger's equation. It seemed to be impossible to apply Schrödinger's equation to the question of whether a *single* electron (or other particle) would pass through one of the slits. With knowledge of the two-slit experiments, we already know that sometimes it will, but sometimes it will not, and if it does pass, we cannot tell the route through which it passed; in fact, we cannot really talk about a route. Schrödinger's equation would indicate in this case that the electron partly passes the barrier and partly does not. This is more or less the case, but there is a catch: Nobody has yet been able to intercept part of an electron! An electron either hits the detector as a whole electron or avoids it completely. Individual electrons could not be considered as waves, because every shred of empirical evidence showed just the contrary.

Thanks to Max Born, this contradiction was resolved as early as 1926, and quantum mechanics became an established and logically consistent branch of physics—even though it has not become more compatible with our worldview on the macro level.

According to Schrödinger's equation, the density of a substance at a given point is given by the square of the local maximal amplitude of the wave function. Thus, if there is a barrier at a certain point, then relying on the square formula, we can calculate the percentage of the electrons that will pass over the barrier. Max Born's great idea was that this does not mean, for instance, that 370 of every 1000 electrons will pass the barrier and the others will not, but that each electron will pass the barrier with probability 37% and will not pass it with probability 63%.

If we think this way, the wave function can be interpreted even for a single electron without being inconsistent with the experimental evidence. As a result of Born's idea, we can arrive after a spot of mathematics at the following conclusion: In the case of a single electron, the square of the amplitude at a given point of the wave function gives the *probability* of finding the electron at that point if we happened to put a detector at that point. This is perfectly consistent with the experimental result that only whole electrons can be measured: It's either there, or it isn't. This *probability of finding* is the physical interpretation of the wave function, and nothing more. And the nature of the electron (and that of other elementary particles) is as it is: In a certain sense it is of wave character, in a certain sense it is of corpuscular character, in a certain sense it has both characters, and in a certain sense it has neither—thus, it is something about which we have no precise picture just yet.

Returning to our amphibious frog analogy we can say that the wave function is invisible at every point of space, but it consists of frogs ready to leap. These are not real frogs, but neither are they non-existent. They are probabilistic frogs. The wave function gives the probability with which a frog will be found at a given point if we look for it there. If we do find it there, then it will always be a whole frog that jumps out, a frog whose size depends only on the wavelength. But we can never tell for certain where it is. No matter where we look for it, we will find it only with a certain probability. The wave function contains only a single frog, but it is located at different points with different probabilities. When we find the frog, it will jump out, and all the other probabilistic frogs will disappear immediately. The wave character will turn into corpuscular character, as we have seen in the two-slit experiment. The wave function is gone, and in its stead there is a real particle.

Today, the great majority of physicists accept the probabilistic interpretation of the wave function. The debate is over the physical interpretation of the frog jumping out of the wave function; or to be more technical: Why, when, and under what circumstances is the *wave function reduced?* We will return to this question in Chapter 11.

Ironically enough, Albert Einstein, the founder of the quantum-mechanical worldview, protested against the probabilistic interpretation until the end of his life, although even he admitted that the model works perfectly, and nobody has found a satisfactory alternative theory.

Thus, the electron plays a kind of mixed strategy in this great game of hide-and-seek. But quantum physicists have found similar mixed strategies in studying other characteristics of the electron (or other particles), for example, their speed of movement. In every mixed strategy the probabilities belonging to a given pure strategy are determined by the wave function. A pure strategy can be one of the possible positions or one of the possible speeds.

We do not know whether Schrödinger named the wave function ψ because he suspected that this idea would have a great effect not only on research into atoms and molecules but also on investigations into the human psyche, or simply because in those days this particular Greek letter had not been appropriated to represent some other important physical quantity. At any rate, as we shall see in the next chapter, certain characteristics of the wave function have proved to be easily generalizable beyond physical systems. Perhaps this is not surprising for a function that included every measurable characteristic of an entity and from which the entire behavior of the entity can be deduced. But let us refrain from hasty analogies: Thus far, no branch of science outside physics and its related fields has been able to use Schrödinger's equation as a technical tool. Until this occurs, let us approach every analogy with healthy mistrust, including my own thoughts on the subject, about which more later.

Chance as an Organizing Force

Since the work of Pascal and Fermat in the seventeenth century, we have had very good mathematical tools for the study of chance. The mathematical field of probability theory has developed powerful theoretical tools. "Theoretical" in the case of probability means, for example, that we assume that a die *really* lands on each of its sides with equal probability, though which side cannot be calculated in advance. According to Newton's worldview there is no theoretical inability to calculate which side. If we knew all the parameters of all the atoms of the die and of the hand that rolls it, and if we possessed the necessary (but hopelessly large) calculating capacity, we could calculate exactly on which side the die would land.

Thus, there is a multitude of *hidden parameters* behind the seemingly chance behavior of the die, the characteristics of all the atoms of the die and those of the person rolling the die, whose values we do not know. If we throw the die in practice, it will really seem to fall on one of its sides perfectly at random, even if according to Newton's physics the result can be calculated—at least theoretically.

According to Max Born's already generally accepted interpretation of quantum physics, the probability waves of the electrons are not based on such hidden parameters. The "position" of an electron is based really on chance, and thus it cannot be calculated, even theoretically. This is why I put the word "position" in quotation marks: We cannot really talk about the position of an electron until we detect it and thus force it to show us its corpuscular character. Perhaps we would like to talk about the position of an electron at a time when we are not detecting it, but that would simply be one of our ill-posed questions, arising from our traditional human concepts, to which nature will not give a sensible answer no matter how much we torment her with our questions. An electron has no location when we do not detect it; in such a case it is merely the sum of probabilistic frogs all over the place.

This would not have bothered Einstein yet. All right, he would say, this is the way we see it now. Some day, as our knowledge of the Old One's secrets grows, we will learn what deeper laws determine this seemingly random behavior. The nature of the debate is revealed in an exchange of letters between Einstein and Max Born. Einstein: "Regarding our scientific expectations, we became opponents. You believe in a God who plays dice, while I believe in perfect laws that rule a world with really existing objects" (*The Correspondence between Albert Einstein and Max and Hedwig Born: 1916–1955*, p. 149). Born replied that if God created the universe with a perfect mechanism, he made at least as much allowance for our imperfect minds that we do not have to solve innumerable differential equations to predict even small parts of it, but He has allowed us to apply the dice quite successfully.

Regarding the role of chance, the opinion of the majority of physicists today is even more radical than that of Born. The work of John von Neumann has greatly contributed to this view. Von Neumann played an important role in developing the mathematics of quantum mechanics. One of his theorems was about so-called hidden parameters. This theorem states that under very general conditions it can be proven that the chance nature of the wave function cannot be caused by things that are not yet known to us, that is, by *hidden parameters* (such as the parameters for the die and hand that, if we knew them, would predict how the die would fall).

In fact, experiments have been designed since then that have shown that the results that arise when there exist hidden parameters (no matter what they are) are calculably different from those when no hidden parameters exist. According to these experiments, hidden parameters do not exist in quantum mechanics. The majority of physicists today believe in the implications of these experiments.

Thus, in all likelihood, the wave function is of probabilistic nature not because our knowledge is limited (even though our knowledge really is highly limited). The probabilistic nature of the wave function arises from the nature of the *universe*. We can say in the spirit of Einstein (although opposing him), "Even if God does play dice, He does so with perfect dice the like of which only He could have created."

Darwin's theory of evolution was the first scientific theory that hypothesized chance as a significant natural force. In Darwin's time the mechanisms of genetics were not yet known, but their discovery has supported Darwin's ideas. According to modern theories of genetics, the particular mixture of genes in an individual is the result of chance, as is the position of an electron when we measure it. Geneticists do not assume hidden parameters, because although perhaps they could design experiments in which the question could be settled positively, assuming random inheritance they have been able to construct a logically complete, consistent, and closed theory, and this theory fits experimental results perfectly. It would not be very promising to employ our knowledge of physics in a search for hidden deterministic parameters behind the highly random mechanisms of heredity. After all, genetic probabilistics could be caused by quantum-mechanical probabilistics, even if genes themselves are a bit too large to be considered quantum-mechanical objects.

But we cannot exclude the possibility that in the case of organisms there may exist yet other sources of randomness. Even if this is so (or perhaps only then), we can nonetheless confirm that actual, ideal chance is a significant guiding principle that plays a role in the world of electrons and genes, and, as we shall see, perhaps even to human thinking.

In game theory the concept of mixed strategy pursues this idea to its logical extreme. It is the essence of mixed strategies that they are based on probabilistic considerations, without hidden deterministic parameters. Otherwise, they would be meaningless, since a perfectly rational opponent with limitless powers of reasoning could calculate exactly what our next move would be. The whole theory of games, including the concept of mixed strategy (or the evolutionarily stable strategy) is meaningful only if the probabilistic outcomes in the optimal mixed strategy really cannot be calculated, not only practically, but *theoretically*.

We know from game theory that the application of a mixed strategy is often the only way of realizing a kind of higher-order rationality, stability, and equilibrium. For example, it is the only way of developing stable biological species. Furthermore, quantum physics has shown that such a model can be applied to inanimate matter, too, even if it should turn out that the issue of hidden parameters cannot be settled conclusively. Game theory has helped us to understand the workings of chance—free of hidden parameters—and how it might function as an organizing principle in the world.

Searching for the Grand Unified Theory

According to Leon Lederman, the *Grand Unified Theory* is the Holy Grail of physics. Such a theory would be a system of ideas (we could also say a system of equations) that consistently and relatively simply would incorporate the descriptions of all the elementary particles and all the physical forces into a logical framework. In the 1980s many first-rate physicists believed that such a uniform theory was within reach.

This optimism was justified by the development of the so-called *standard model* of quantum physics. This model summarized all known particles and forces (with the exception of gravity) within a rather uniform framework. Physicists have generally become more pessimistic in recent years, although the standard model has held up well. For example, with the aid of a kilometers-long particle accelerator constructed for barely a few hundred million dollars, researchers found a particle called the *top quark* that had been predicted by the model in 1994. The standard model includes almost everything. In fact, only one type of force—a very weak force in comparison to the forces within the atom—sticks out painfully, namely, gravity. The other greater or lesser anomalies will certainly be resolved by a few garden-variety Nobel Prize–winning discoveries.

Einstein was searching for a relationship between gravity and the electromagnetic forces of molecules already in 1901, at the age of 22, but he searched for it unsuccessfully until his death. In 1915, he succeeded in logically proving the existence of gravitational fields within the general theory of relativity, which includes a built-in symmetry from which this type of force arises and that is different from electromagnetic forces. However, Einstein did not believe that the road to unification leads through quantum theory. Today, it seems almost certain that it does.

Lederman remembers a lecture in the 1950s in which the physicists Werner Heisenberg and Wolfgang Pauli spoke about their idea of a uniform theory of elementary particles: "Pauli's final comment was an admission: Yah, this is a crazy theory." Then Niels Bohr made a remark that has become a common saying since then: "The trouble with this theory is that it isn't crazy enough." As so often in the history of quantum physics, Bohr was proved correct: The theory in question, and a dozen others, long ago vanished from the scene.

The chief problem with this unification theory is that the mathematical description of the reduction of the wave function (the probabilistic frogs jumping out as real frogs) incidentally contains a certain mathematical description of the geometric structure of space, a different structure from that given in general relativity theory. Mathematically, both types of geometry are perfectly logical and consistent, but they are mutually exclusive. Since quantum theory and relativity theory work with objects of very different sizes, this incompatibility never causes practical problems: Each theory works when applied to the appropriate objects of investigation. But it is not very elegant or satisfying that two individually excellent theories cannot be reconciled. Who knows what we would be sweeping under the rug if we ignored the problem with an, "After all, the two theories together meet every practical requirement."

But the question is interesting not only from an aesthetic aspect. We already know one point of intersection between the two theories: cosmology. Cosmology studies the origins of the universe. Current cosmological theories (at least those based on the principles of physics) agree in the so-called big bang theory, which states that the universe began with a violent explosion. Quantum theory and relativity theory differ so much in their domains of inquiry that every question about what happened after the first few seconds of the big bang can be answered quite reassuringly. But the beginning of the beginning, the physical events of the first few seconds, remain obscure because of the incompatibility between the geometries of the two theories.

The theoretical physicist and mathematician Roger Penrose wrote in 1989 (for those readers who find the following incomprehensible, an explanation follows), "My own point of view is that as soon as a 'significant' amount of space–time curvature is introduced, the rules of quantum linear superposition must fail. It is here that the complex-amplitude superpositions of potentially alternative states become re-placed by probability-weighted actual alternatives—and one of the alternatives indeed *actually* takes place."

This is the idea that connects quantum physics with game theory. Essentially, Penrose is saying that the description of the "probabilis-tic frogs" by today's standard quantum mechanics is probably too complicated. In exchange, the technical apparatus with which we can do actual calculations is relatively simple. Probably, the desired unified theory will have to content itself with a simplified descrip-tion, similar to the way we introduced the nature of the probabilistic frogs: They are simply at one place with a certain probability and at another place with another probability. They move at a certain speed with a certain probability and move at another speed with another probability. Undoubtedly, we shall have to pay a price for this simpler description: The technical apparatus will become more complex, but we cannot tell yet how complicated and in what ways.

In the probabilistic description of the mathematical apparatus of present-day standard quantum mechanics a kind of geometry is au-tomatically built in. This geometry describes the structure of space slightly imprecisely on the macro level. But in order to avoid this im-precision we would have to throw out quantum physics, which is a complete theory according to our present knowledge and which works excellently. Moreover, no alternative theory of similar value is on the horizon.

Mathematics cannot do more than point out logical equivalence or inconsistencies among several possible geometrical worlds (for ex-ample, the geometries of Euclid and of Bolyai and Lobachevsky). It is the job of physicists to decide what the world is really like. However, according to our best present theories, the geometry of the small and the geometry of the large are different. The Grand Unified Theory that would bridge this difference has yet to be hatched.

In the past century modern physics has radically changed our conception of time and determinism, ideas that had been relatively unchanged for thousands of years. Perhaps only our fundamental view of the geometry of space has been left more or less untouched (the slight curvature of space has not altered our attitude significantly). It may well be that the Grand Unified Theory, if and when it comes, will fundamentally upset even this concept.

According to Penrose's intuition, we might be able to solve our current impasse if we threw out the geometric excess from the quantum-physical description. After all, particles are just particles. What do they have to do with the geometry of the space they live in? What would remain would be real alternatives with their accompanying probabilities; or in the terminology of game theory, the pure strategies of the particle and the probabilities chosen by the particle according to which it implements these strategies—in other words, the mixed strategies of the particles, without hidden parameters and geometry, which is just how game theory treats mixed strategies.

The Great Game of Nature

From the perspective of game theory, it is no wonder that elementary particles realize mixed strategies, since we know that in certain games this is the only mechanism for achieving stability. Of course, "particle" is only the name of whatever it is that we call a particle playing a mixed strategy, since the meaning of the word has changed. We have learned that a particle has a probabilistic character, and in order to fulfill its potential, it acts as a probabilistic wave. In short, it is as it is. It exists. After all, what's in a name?

With this view, however, we may now inquire into the nature of the game in which elementary particles take part. What are the rules to which they conform when they choose their mixed strategies? Can we say in any sense that the elementary particles play optimal mixed strategies? What, in sum, are the fundamental principles of Nature's great game?

It may well be that the birth of the Grand Unified Theory will provide an answer to these questions as a byproduct. But we have another question on the table, one that resonates with these questions of physics: What might be the operating principles of the natural force called evolution? The connection between the two questions is that both are related to basic units that have no further internal structure—at least from the perspective of their respective branches of science. One gene is like another: If a gene is exchanged for an identical copy of the same gene from somewhere else, nothing will change. The same is true of elementary particles: If an electron from a horse is exchanged for an electron from a horseshoe, no change will occur either in the horse or in the horseshoe. This is a logical consequence of quantum mechanics.

As we saw in Chapter 8, evolution uses mixed strategies, and moreover, it mixes the *principles* of gene selection and group selection. These two rational principles affect species in different proportions. Elementary particles also mix their pure strategies: They can assume their various states with certain probabilities. We do not yet understand the guiding principle according to which these probabilities are selected. The existence of such principles, however, is not unknown to physics.

Newton's mechanics can be fully constructed relying on the so-called *principle of least action*, a sort of Occam's razor, according to which an object moving freely from one point to another will always choose a course that requires the least amount of energy, thus optimizing its path. Thus put, this principle is strange, because it is perfectly teleological to assert that the movement of an object is governed by the goal the object seeks to attain. Strangest of all is that Newton's laws can be deduced from the principle of smallest effect, and conversely, the principle of least action can be deduced from Newton's laws.

Thus a physics that is based on a principle according to which every physical system strives to attain some goal is *mathematically equivalent* to a physics in which no such assumptions are made. Physicists have swept this strange fact under the carpet just as most evolutionary biologists since Darwin have regarded evolution as goalless, since the assumption of a goal is unnecessary. Nevertheless, an unknown goal may exist, and its existence may arise from goalless laws of evolution (or from Newton's equations, which also do not assume a goal). This is why we could say earlier that banishing the concept of meaningfulness from biology does not exclude its later emergence, even if the theory of evolution proves to be correct as currently formulated.

This is the sense in which a higher-order principle may govern the independent movement of elementary particles or the natural selection of species. It is possible that elementary particles and genes realize an as yet unknown kind of *optimal* mixed strategy that brings about stability and equilibrium among the participants in nature's grand game. Perhaps John von Neumann found an essential component of the future Grand Unified Theory in his theory of games, while it has already been proven that he laid the mathematical foundations of two different theories of evolution. What, then, are the rules of nature's grand game? What are the guiding principles of the players? What are the general laws of the mixed strategies applied by evolution and those played by elementary particles?

Even if we do not yet know the answers to these questions, we can consider them within the framework of a uniform, effective system of ideas—the concepts of game theory, which have exposed sources of diversity through the concept of mixed strategy from the level of elementary particles to biological evolution. We may hope that this will lead to important knowledge about human thinking and about the reasons for and meaning of the diversity of human thought. After all, game theory was originally about human thinking, and it may prove to be a guiding principle in the study of rationality.

THE PSYCHOLOGY
OF RATIONALITY

11

Loves Me, Loves Me Not . . .

What a girl plucking the petals from a daisy really wants to know is whether she herself is in love.

No matter how we define rationality, human thinking does not employ completely rational means in going about its business. Man does not even solve with pure logic those problems for which logic would be a perfectly suitable and very effective method. I demonstrated quite a few examples of this in my book *Ways of Thinking*. Yet we are sometimes able to carry out logical, syllogistic reasoning, and despite our fundamentally illogical nature, we arrive at correct solutions. The duality of rational and nonrational thinking is a great mystery in the psychology of human thought.

Game theory has created a higher-order system of thought in which the paradox of this duality can be resolved, at least to a great extent. The royal road, if such there be, of how it accomplished this leads through the world of quantum mechanics. (Let us keep in mind, however, the words of Archimedes to King Hiero—or, following Proclus, those of Euclid to Ptolemy I—when the latter demanded a quick and painless initiation into the mysteries of the mathematician's art: "There is no royal road to mathematics.")

Schrödinger's Cat

We mentioned in Chapter 10 that the probabilistic interpretation of the wave function has been accepted by the majority of physicists. Only a few physicists would object to our analogy that the wave function describing the electron or other elementary particle behaves essentially like a "probabilistic frog." The professional debates are mostly about the physical meaning of the frog jumping out of the wave function, or, more technically, Why, when, and under what conditions is the *wave function reduced*? Why do the probabilistic frogs of the wave function change into a single real frog? How does the photon (or electron) "know" that it has to behave as a particle when it is observed as a particle? Why doesn't the wave function avoid the detector, just as it avoided the lead screen and passed through the small slit?

The greatest problem is that if we do not observe the electron at the detector (say we forget to turn on the detector), the electron will continue to avoid the detector just as it avoided the lead screen! If we nevertheless observe it at another detector wall, we'll find that the first row of detectors has affected the electron just as much as the screen with the two slits: The first row of (inactive) detectors maintained the electron as a wave function, that is, as a totality of probabilistic frogs, although divided according to the number of slits. Many excellent physicists have raised the idea that perhaps *conscious human observation* itself has led to the reduction of the wave function and to the jumping out of the single real frog. But this sounds most improbable. Could our consciousness really influence the physical world? And if yes, whose consciousness?

Schrödinger illustrated this dilemma with the help of a cat (rather than a frog), and this cat has become a favorite topic of discussion among physicists and laymen pondering the secrets of quantum physics. Here is the bittersweet story of Schrödinger's cat: A cat is placed in a box with solid walls. In addition to the cat, there is a piece of radium in the box that emits a decay particle every hour with 50% probability. This is a pure probability, free of hidden parameters. Thus, the time of the decay of the radium cannot be predicted theoretically, not even if we knew all the parameters of all of its atoms. If a particle is emitted, a detector within the box detects it immediately, opens a valve, and lets some cyanogen into the box, killing the cat. An hour later we open the box to see whether or not the cat is still alive. Schrödinger's question is this: What can we say at the moment before opening the box? Is there a living cat or a dead cat in the box? Well, at this moment we cannot say whether the cat is dead or alive. The best we can say is that the cat is alive with probability 50% and that it is dead with probability 50%. We can say no more about the condition of the cat as about the position of the electron before we detect it in the two-slit experiment. When we open the box, we can say with 100% certainty that the cat is alive or with 100% certainty that the cat is dead. No matter what we might say beforehand about the condition of the cat, it would be only a statement of probability, because there is a real source of chance in the box, free of hidden parameters, whose behavior we can neither guess nor calculate.

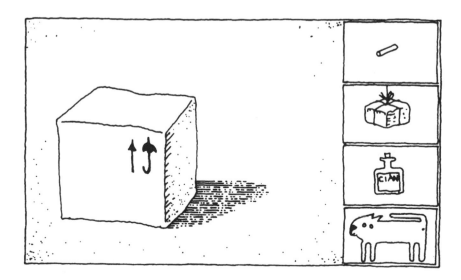

If when we open the box we find a dead cat, can we say that the death of the cat was a reality before the box was opened? Had we not opened the box, would the cat have continued to live happily with 50% probability? In any case, the box as a system would have behaved exactly as if there were a cat living inside with a probability of 50%. Is it possible that the observation itself killed the cat or, if the cat is alive, that observation brought the cat to life from a probabilistically half-dead state? Perhaps that would be going too far, but undoubtedly, the system is in a state where we *have the possibility* that our talking about the life or death of the cat with 100% certainty is a result of conscious observation.

Digression: "Poetic" Thoughts

There are many views about the relationship between quantum mechanics and consciousness, as many as there are writers on the subject, but no reassuring answer has yet been found. The differences between the possible interpretations are substantial. To give an idea of the range of opinion, let's look briefly at interpretations given by a few physicists. I shall try to distill them into a sentence or two—they will look more poetic than scientific this way. It will be their common characteristic that none of them contradicts the *equations* of quantum physics, but there is as yet no chance of proving any of them experimentally.

According to Jenő Wigner, a completely different physics may apply to conscious beings, a physics whose fundamental laws differ from any now known. Roger Penrose and Ilya Prigogine think that there exists a radically new physics between the micro and macro levels, whose laws differ from the other two levels, and we have not taken even the first step in discovering this level—such a physics might explain consciousness. According to Danah Zohar, the particle–wave duality corresponds exactly to the mind–body duality. Therefore, a special theory has to be developed for the physics of the observer. According to John A. Wheeler's "participating universe" theory, our past is only a potential existence; our present consciousness changes our past into real existence. According to Paul Davies, our consciousness forms a randomly branching wave function at every moment, and actually it is not God, but we ourselves who keep rolling the dice. Hugh Everett suggests an interpretation according to which the world is branching into an infinite number of copies every moment, and we are actually present in an infinite number of copies, with an infinite number of life histories, in an infinite number of equally existing worlds.

So much for poetic obscurantism. The world is as it is, and quantum mechanics is nothing but a theory created by humans to describe certain phenomena. This theory works with astonishing precision. For instance, we can build microprocessors and nuclear power plants with its aid, but like every theory, this one, too, has its limitations.

My little summaries have been unjust to the highly respected thinkers whom I have cited, for every soft idea has a hard, technical core that I did not even touch upon, a core that proves mathematically that these ideas do not clash with the equations of quantum mechanics. Thus, quantum mechanics does not exclude the possibility that the world is consistent with any of the interpretations, even if some of the interpretations exclude one another. Naturally, it may turn out that none of them encompasses the true nature of the universe. But it is also possible that several of these interpretations will prove to be true in some deep way.

It is a common feature of the above interpretations (except that of Everett) that they look for the resolution of the paradoxes of quantum mechanics in consciousness, but none of them considers consciousness necessarily as an entity based on the principle of pure rationality. In fact, according to Penrose, for instance, the new physical laws he hypothesizes will entail not only that the functioning of consciousness not be purely rational, but that it have clearly nonalgorithmic components as well. Thus, it cannot be modeled by computers—by Turing machines—even theoretically, not even if we build in a perfectly random random-number generator.

Although these poetic thoughts are all the ideas of outstanding physicists, the majority of physicists today think that consciousness does not have to be introduced in solving the internal problems of physics. For example, Wheeler, whom we cited above, expressed such an opinion in a later paper, writing that the reduction of the wave function might as well be caused by the fact of observation itself, regardless of the conscious nature of observation.

Jenő Wigner also long thought that consciousness causes the reduction of the wave function, but his opinion changed later in his life. In his old age he said that it is self-conceit of humans to think that consciousness *causes* such phenomena. What we do not know, we do not know, and that's it. Quantum mechanics says nothing about consciousness. Bohr and Heisenberg also held this opinion.

Let us end our "poetic digression" with some thoughts of Jenő Wigner's: "The miracle of the appropriateness of the language of mathematics for the formulation of the laws of physics is a wonderful gift, which we neither understand nor deserve. We should be grateful for it and hope that it will remain valid in future research and that it will extend, for better or worse, to our pleasure, even though perhaps also to our bafflement, to wide branches of learning."

The Contingency of Human Concepts

Schrödinger's thought-experiment can easily be misunderstood if we take it literally and think of a real box containing a real cat. The box is a relatively exact analogy of particles as conceived by quantum mechanics only if we consider the whole box—together with the cat, the radium, and the cyanogen—as a single system that cannot be divided further (that is, if it is conceived as an elementary particle). Thus, the box *together with its contents* corresponds to the electron or photon; the cat and the radium cannot be separated from the box. This does not mean that the box cannot have an internal structure (as in our example it did have), just as elementary particles also may have their own internal structures, including their own force fields and their own complex wave functions. But elementary particles are elementary for the very reason that this internal structure cannot be subdivided into smaller components of matter. Rather, the internal structure characterizes the elementary particle in its totality. In our analogy, this complex internal structure was represented by the unity of the cat, the radium, the particle-detector, and the cyanogen. Do not be misled by the fact that a real cat or piece of radium consists of a number of atoms, electrons, and other particles. In the *abstract box* of Schrödinger's cat such things no longer exist.

But how can we talk about matters of life and death in the case of such a cat, a cat that can no longer be divided into smaller parts? We can talk about these things just as much—no more, no less—as we can about position in the case of an electron! Life, like location or velocity, is a *human* concept. These are the concepts we have developed at our present level of thinking and for the perception of which we have relatively reliable tools. The nature of Schrödinger's cat is not life or death, but living or not living with a certain probability. The nature of the electron is not being here or there, but being simultaneously at different places with certain probabilities.

The opening of the box corresponds to the act of measurement and observation. Thus, this is the moment when the system *cannot do otherwise* than to change its *mixed state* (in which the cat was alive or dead with a certain probability) to a *pure state* and show us either a definitely living cat or a definitely dead cat. The box–cat system would never do such a thing of its own accord. Without external intervention, the box would go on forever in a state in which a cat is living inside with a certain probability—although this probability would gradually decrease over time. Measurement according to our human concepts forced the box to reveal to us for a moment the cat and to produce an actual life-or-death phenomenon—just as our measuring device required the electron to show us momentarily a real location.

Imagine what would happen if after having opened the box and determining that there is a dead cat within we resealed the box, with the radium and detector exactly as before. Apparently, this "reinitialized" box could not be the same as it was before we opened it, since we now know that there is a dead cat inside. However, if we want to symbolize the behavior of elementary particles with Schrödinger's cat, we must say that *the cat in this reinitialized box will continue to live probabilistically*. Thus it is possible that when we open the box later, we'll find a living cat within! If not cats, at least electrons do something like this. After its detection it continues its existence as a sum of "probabilistic frogs" as if we had just fired the electron gun at the detector. It is quite clear from this how arbitrary are all our human analogies regarding the quantum world. Unless one believes in metempsychosis, our analogy fails at this point. (It just so happens that I do not believe in reincarnation. I didn't even believe in it in my previous life.)

As with every analogy that we create, Schrödinger's cat is really not suited to describe the behavior of elementary particles, but it can help in understanding how "probabilistic frogs" can work, how a system can be in different states simultaneously, and how it can nevertheless take on an unambiguous state as a result of measurement and observation.

In terms of game theory, we can say that the cat plays a mixed strategy. To be more precise, it is not the cat that plays the mixed strategy, but the whole box, of which the cat is an inseparable part. It is the box that reveals a perfectly live or definitely dead cat upon being opened, but life and death are meaningful concepts only in our clumsy human conceptual framework. For the box, they have no meaning at all.

Stone–Paper–Scissors Redux

Let us return to the player who, while we have been talking about other things, has all the while been playing the stone–paper–scissors game strictly according to the optimal mixed strategy. This player based his decision strictly on blind chance. He shows stone with probability 1/3, paper with probability 1/3, and scissors with probability 1/3, though he uses a die rather than a piece of radium to decide on his move, a method perfectly suited to his aims on a human scale.

As we saw in Chapter 10, Schrödinger used the symbol ψ to describe the complete state of any physical system. This physical system could be an electron or even a larger object. Now let this larger object be the person playing the stone–paper–scissors game. We could perhaps say the person's *psyche*, but as we stated in Chapter 5, the odd thing about playing an optimal mixed strategy is that in a purely psychological game the player may completely disregard psychology.

In the case of Schrödinger's cat, ψ represents the whole box, which is in one of two pure states: The cat inside is either alive or dead. These two states exist as a mixture of probabilities until we open the box, that is, until we observe the system. The fact of observation brought ψ into a pure state.

In the case of the individual playing stone–paper–scissors, it is the player who is the entity ψ, which now has three possible pure states: stone, paper, scissors. More precisely, the player is in, say, a "stone" state if he has already decided to show stone on the next move. The player playing a mixed strategy is in a mixed state until he rolls the die just as much as Schrödinger's cat is in a mixed state in the box until it is observed. An optimal player will look at the die only at the moment when he is commanded to show his hand, and he will then show immediately what is prescribed by the die, for this way the opponent will have the least chance of guessing what he will show. At that moment the player has been transformed from a mixed state to a pure state.

Perhaps the analogy of the player playing stone–paper–scissors illustrates the fundamental principles of quantum mechanics better than Schrödinger's cat is able to do. In the case of the player, we can conceive not only the essence of the mixed state and why he enters into a pure state upon observation, but also the fact that the player returns to a mixed state immediately after observation. For after the observation, the player does not know what he will show the next time until the next command is ordered and the die is rolled, that is, until the next observation takes place. The player is in a stable mixed state between two commands just as much as is the electron between two observations.

The Cheating Innkeeper

The player who makes his decision on the basis of rolling a die in the stone–paper–scissors game can be considered completely rational. He knows game theory, and thus he knows that unless he has reason to assume that he is smarter than his opponent, the wisest course is to follow the optimal mixed strategy. This player acts exactly according to the principle of rationality, provided that he has calculated correctly the probabilities of the individual pure strategies, prepared a fair die with the appropriate number of sides, and really obeys the result dictated once the die is cast, in short, if he really does everything that is expected of the rational Martian in Chapter 2.

This idealized player stands completely outside the interest of psychology. His actions are not motivated by psychological forces, but by purely mathematical principles. The way he plays is completely automatic. In fact, his play can be automated. After the conscious decision to follow the optimal mixed strategy has been made, the player employs no conscious or unconscious elements. For the duration of the game, let us not even consider the player to be a human being. At times like this, he is no different from an electron.

In order for our idealized player to begin his optimal strategy, however, he must be fully aware of the values of the different outcomes of the game. This is not always an easy task in the real games of life. It is easy for an electron, however. The laws of physics determine its mixed strategy at every moment, even though we, human beings, do not know these laws yet. The mixed strategy to be followed by man, however, is greatly determined by his own subjective values. As we saw in Chapter 4, a different set of values may radically change the optimal strategy of a game. For instance, choosing the golden rule can transform the awkward situation of the prisoner's dilemma into an extremely simple decision.

Human decision-making is usually governed by several complex elements operating in concert. It is very rare that we have to choose between being given $100 and being given $200, while all other conditions remain fixed. Usually, we can study a situation from the points of view of economics, ethics, utilitarianism, ecology, culture, etc., and the results are often contradictory. Human beings are not governed by universal laws as are elementary particles. The mixed strategies of people are guided by many aspects more or less particular to them. Practically, they cannot be quantified.

All this, however, does not change the fact that even an approximately good decision can be made only using mixed strategies. Man is forced to figure out somehow the subjective values belonging to the different outcomes of a situation (or game) and to determine the probabilities for the different decisions (or moves, to employ the game-theoretical term). These two tasks are usually not solved by people in the manner suggested by game theory, according to which first the values are determined and then the probabilities are calculated. The two phases are usually merged in human thinking, and then a decision pops out as the end result. We are far indeed from making our decisions by purely rational means.

There is a story, I vaguely remember reading it somewhere in Rabelais, about an innkeeper who, when he has a difficult decision to make, rolls the dice to determine how he should act. But since he is at heart a swindler, when he is rolling the dice he always does his best to cheat. Yet he obeys the dice unconditionally.

Evidently, this innkeeper does not act according to the rules of pure rationality, but what he does is not necessarily irrational. If he cheats well, he will make good decisions remarkably often. Thus, the procedure of this Rabelaisian innkeeper cannot be considered purely rational or purely irrational. If something is not rational, it does not mean that it is necessarily irrational or unreasonable.

Quasi-rationality

The pining shepherdess who plucks petals from a daisy to gauge the constancy of her swain does not take the course of pure rationality. Yet it is also possible that her action should not be considered irrational. In the course of pulling the petals she may acquire knowledge that throws light on the *real* truth of her situation; this truth may not have been available to her by the methods of pure reason, by pure rationality. If this is really the case, then pulling the petals cannot be considered a priori an irrational act.

We are going to call acts, patterns of thought, and evaluations of situations that do not follow the rules of some pure rationality, yet do not contradict them either, as *quasi-rational*, differentiating them form irrational acts—those that contradict reason.

On occasion, even irrational reasoning may lead to truth. For example, if I say $2 + 2 = 6$ and $6 + 4 = 8$ and then substitute the 6 in the second equation by the $2 + 2$ of the first equation, I will be correct in saying that $2 + 2 + 4 = 8$. However, irrationality, as opposed to quasi-rationality, can hardly be expected to lead to deep, *general* truths.

Since Rabelais's innkeeper does not know the real values of the outcomes of his various options, it would be hopeless from the outset for him to attempt to determine the probabilities of his optimal

mixed strategy. Nevertheless, the procedure of the innkeeper, including his cheating, may have a deep meaning. He makes his decisions
on the basis of rolling the dice, and thus he plays a mixed strategy, as
prescribed by pure rationality. Yet he also cheats, by which he lets the
intrapsychic forces he does not know well have an effect on him.
These will determine how much weight he gives to his cheating, that
is, how much chance he really should give to the less desired, but
somehow not dismissible, options. Thus may a practiced swindler be
able to approach the unknown and indeterminable optimal mixed
strategy. Thus, we are justified in calling his procedure quasi-rational.

Plucking Petals

Plucking petals from a daisy seems even less reasonable than the
innkeeper's dice-rolling. Plucking petals can in no way be considered
an act of observation by which our shepherdess brings her beloved
into a state of pure strategy, thereby resolving the troubling question,
Does he love me? Furthermore, if she pulls out the petals in solitude,
plucking the petals will make him love her no more and no less. But
this is not the true meaning of the game.

The point of this line of reasoning has been revealed in the epigraph to this chapter. What our petal-plucking shepherdess really
wants to know is whether she herself is in love. In all probability the
thought process leading to this conclusion is more interesting than
the conclusion itself.

First, let's have a look at the most obvious counterargument: Why
do we maintain that the shepherdess doesn't really want to know
whether she is loved by her swain? On the one hand, why shouldn't
she? And on the other, she may also need to know this for practical
purposes: She will need a different strategy to win him if he loves her
but has been too shy to reveal his love than if he has never considered her a possible object of his affections.

If she is sure of her feelings, then she is also sure of her goal: to
win the object of her love, to elicit love in return. In choosing the
best method of attaining this goal she may find it useful to know
whether or not he loves her, and it is entirely possible that she can

find no better method of finding this out than to pluck petals from a daisy. But in turning to such a seemingly irrational method, our shepherdess might well build her goal into it, turning to sorcery to secure the object of her love. Should she tie catnip into her hair? Or mix with mud a pregnant rabbit's urine and smear it on the sole of her love's shoe? Should she recite incantatory verses?

The most rational strategy of this game is most probably some kind of mixed strategy, and so a random-number generator, such as whether the number of petals is odd or even, may not be out of place. But this does not seem sufficient explanation for why the loves-me-loves-me-not game is so widespread. Furthermore, the parity of a flower's petals is usually a poor random-number generator, since most flower species have a characteristic odd or even number of petals. Nevertheless, the outcome is still uncertain, because flowers often lose a petal or two. Experienced loves-me-loves-me-not players already cheat a little in choosing the flower itself, just like Rabelais's innkeeper, and select a flower from a species characterized by an odd number of petals. However, those who know exactly what their aim is rarely waste energy on such childish games, for there are more direct methods available to them, though these methods are not necessarily more rational and may also require mixed strategies.

But such a universal tradition most likely has a more constructive goal than merely finding out whether our love loves us. Magic, even at its most harmless (like petal pulling), is basically constructive. Let's see what we get if we approach this simple game from the viewpoint of quantum mechanics. The analogy between elementary particles, Schrödinger's cat, and the loves-me-loves-me-not game can be summarized in the following table:

	elementary particles	Schrödinger's cat	Loves-me–loves-me-not
the system (ψ)	electron, photon, etc.	the box with all its contents	the psyche of the musing person
the human concept	place, speed, etc.	the life or death of the cat	the blossoming or extinguishment of love
the reason for reduction of the wave function	observation of the particle with an intervening detector	opening the box (intervention for the sake of observation)	the musing person's own consciousness as an intervening external observer

We have already discussed the analogy between the first two columns. The analogy between the second and third columns is striking, but what justified our setting up the third column in the first place? And why did we choose these particular concepts? If the game loves-me-loves-me-not is considered from the perspective of its original, literal, sense, then the observation made by the player is meaningless from the "perspective of quantum mechanics," since it does not bring the observed object into a pure state. We have suggested, however, that the real aim of the game might be something else. This suspicion has been formulated in the third column of the table: If pulling the petals now becomes meaningful from a quantum-mechanical perspective, then our suspicion is greatly reinforced. This kind of thinking is perhaps unusual in psychology, but such a strong relationship with a logically clear scientific theory is definitely worthy of attention.

In order for the third column to be reasonable (now from a psychological perspective), we have to answer at least three questions. (1) Why do we need to bring our own psyche from a mixed state to a pure state? (2) Why do we need to pull out petals to observe our own psyche? Why can we not observe it directly, in which case we would we able to understand the reduction of the wave function without further ado if that is what we wanted. (3) If this is all so, why do we murmur "loves-me-loves-me-not" while pulling the petals instead of saying "I-love-I-love-not"?

My answer to the first question is that the problem lies not in the difficulty of attaining our goal but in the lack of clarity as to what our goal really is. The determination that we love the other will greatly influence our further course of action. We will no longer have to wonder what aims our actions should serve. It will suffice to look for actions that best serve our known purposes. This is why it is useful to bring our psyche into a pure state, at least for a moment. Besides, it also follows from our analogy that our psyche will again be in a mixed state right after pulling out the last petal. However, when we experience a pure state momentarily, this moment of clarity may determine and give meaning to our actions for a long time to come.

The answers to the second and third questions come from similar considerations. A key word, one that plays an important role in a number of psychological techniques, is *distancing*.

Distancing

The human psyche is naturally in a mixed state, just like elementary particles. Yet, in order to make decisions and take action, we usually have to choose a pure strategy. Sometimes we need to do this simply because the current mixed state creates a degree of tension in us that is difficult to tolerate. Thus, there may be several good reasons for forcing ourselves into a pure state. According to our analogy, this can be achieved only by an external observer, indeed, in the view of many physicists, only by a *conscious* observer.

In standard psychotherapeutic treatment, or in hypnosis, the role of the conscious observer is taken by another person, one who knows the techniques of observation—rather like the physicist observing electrons who knows how to employ the detecting device. But now we are interested in the strange situation when the observation is made by a person on himself, being thus simultaneously the observer and the observed.

It is customary to interpret the example of Schrödinger's cat as a real cat who can even be a conscious observer—at least while it lives. We have considered the cat as an integral part of a box that has no further elementary particles. This was so we could make the analogy with real elementary particles more exact. At the same time, the other side of the analogy—with our shepherdess plucking petals— becomes less exact.

Schrödinger's cat is only an analogy. The human psyche, however, has a particular characteristic that elementary particles do not: It is conscious of itself. Thus, theoretically, it is capable of observation. But this seems to be a contradiction. On the one hand, we have said that the psyche's natural state of existence is a mixed state, which would never change into a pure state of its own accord; but on the other hand, it is still able to bring itself to a pure state by its own conscious observation. We are not yet able to resolve this paradox at the *theoretical level*. In *practice*, however, we know some psychological techniques by which we can realize this even in difficult situations. One of them is plucking petals. In the next chapter we are going to discuss some techniques that are more complex and more general.

As a practical tool, the flower with its petals serves the purpose of distancing: to separate the psyche in its natural mixed state from the consciously observing psyche as much as possible, and to enable these two entities to function simultaneously. The aim is for the latter to bring the former into a pure state, just as the electron detector forces the electron, whose natural mode of existence is a mixed state, to switch to a pure state.

The flower also serves the purpose of distancing if the shepherdess plucking the petals does not talk about herself, but about another person, even if it is the object of her love. Pronouncing *I* is an ancient taboo, which, according to psychologist Judit Kádár, was perhaps first broken by Oedipus when he solved the riddle of the sphinx ("In the morning it walks on four legs, at noon on two, in the evening on three—what is it?" The solution: man). Today it seems incredible that this simple puzzle was so difficult that an Oedipus was needed to solve it and that it could give the sphinx such power over the city of Thebes for so long. However, if we realize that for the solution of this innocent and—from the aspect of "puzzle-ology"—totally fair puzzle an ancient taboo had to be done away with, then everything becomes understandable. Although Oedipus did not have to pronounce the forbidden word—he had only to say, "It is man"—nonetheless, the taboo so restricted the thinking of the citizens of Thebes that despite their motivation to solve the riddle they could not find the solution.

Depersonalization is another important technique of distancing. In meditation, for instance, it helps to bring about a pure state of consciousness if we imagine something we want to know about projected on a screen. This is also a common technique in the practice of hypnosis. If we imagined the thing we wished to know about directly, then our procedure would be totally rational. Experience shows, however, that if the phenomenon to be thought over affects us very strongly emotionally, we will usually not be able to do this. Our mixed state often remains unresolved by our direct, logical thinking. And what is worse, logic often makes us even more confused than before.

Meditative cognition, on the other hand, may be greatly helped by the distancing technique of projection onto a screen. Projection onto an imaginary screen, the simple observation of the behavior of the projected person on the screen without the use of logic, is a quasi-rational procedure. No rational motivation, no logical law, forces us to do this, but neither can it be considered irrational.

12

Rational Irrationality

My brain understands it, but I don't.

When plucking petals, we often try to cheat a little, cleverly, so that not even *we* notice the deception. In a moment when no one is looking—ourselves included—we take off two petals "accidentally"; or when a few petals still remain we toss away the flower—just "in time"—saying that the flower was irregular, that it was defective, and therefore we have to begin over with a new flower. The only judge is the psyche of the person in love; the very psyche who decided to cheat. Therefore, it is probable that the judge will be lenient, at least to a certain extent, for example by permitting another try. We often thus regulate the probabilities associated with a mixed strategy, even if we do not take off real petals or roll real dice, but simply ponder our own affairs. Sometimes we accept the result of our first happenstance train of thought, while at other times we try to think things over from other perspectives as well.

Man realizes by such small "cheats" what our Martians in Chapter 2 achieved by a certain mathematical competence. In this way man appropriates to himself the right of throwing the dice as often as necessary to reflect the current mixed state of his psyche.

Correct Decisions Reached by Inadequate Methods

The lover (or nonlover?) who begins to take off the petals again echoes the technique of Rabelais's innkeeper: *He cheats in the interest of justice.* Our shepherdess would like to bring her own psyche, which is in a mixed state, into a pure state, and to know at least for a moment whether or not she is in love. She wishes to reach this pure state so that the opposing forces working within her will reveal their will. If, for instance, she is more in love than not in love, the observation will give the result of "loves me" with a higher probability. However, at the moment only the flower is available to her, a flower that she believes has an odd or even number of petals with equal probability.

This *unjust* equality of probabilities has to be compensated somehow. We would cheat ourselves if we did *not* compensate, for we would give an unjust advantage to the state having the lower probability. Perhaps this is why we need such a complex rite as plucking petals instead of, say, tossing a coin. This way the psyche has world enough and time to map out the real probabilities. In the meantime, it can resort to greater or lesser tricks, or as a last resort, it can declare the whole game rubbish.

The innkeeper's method of decision-making, including his trickery, was more exact than plucking petals with cheating, and moreover, he took the result seriously. The innkeeper could mobilize one of his *very highly developed abilities* in the interest of a given aim, and that is why he could allow himself to base his ultimate decision on the roll of the dice and to obey their order unconditionally. Of course, the innkeeper did not cultivate his abilities as a swindler for the purpose of resolving dilemmas. Nevertheless, he developed a refined and complex ability that was able to serve him better in his decision-making through its complexity than a perhaps more suitable, but less cultivated, method—like game theory, which in any case was completely unknown to him—would have done.

This is how most human thinking functions. It makes important decisions by totally inadequate, but from another aspect highly refined, methods, and these important decisions prove by and large to be good ones, especially if we consider that it is often the case that no absolutely correct decision exists. Game theory has shown that *theoretically* the most correct decision is to follow an optimal *mixed* strategy. Thus, when we talk about somebody's decisions as being correct by and large, we should not judge the correctness of individual decisions, but that of the totality of the decisions. Do the decisions appear in proportions commensurate with what would be observed if an optimal mixed strategy were being followed?

Maybe we are too optimistic in believing that human decisions are by and large good ones. For the time being, we have two arguments in favor of such a belief. Our first argument is based on the fact that mankind, despite its frail constitution, has shown itself to rank among the fit species. It could not have made wrong decisions too often, because then natural selection would have extinguished it. Our second argument is that most of our decisions are actually more or less automatic, and no long cogitations are necessary even if the task is very complicated. Most of the time, we make the correct decision about how to hit an approaching tennis ball so that it goes back over the net. Theoretically, if we wanted to take into consideration the velocity, trajectory, and spin of the ball, and many other important factors such as the wind speed and the tension of the strings on our racquet, we would have to be able to solve a system of complex differential equations in order to hit the ball at all. Most of the better tennis players are not more knowledgeable about differential equations than the average duffer, just as Rabelais's innkeeper did not know game theory. Tennis players use theoretically inadequate, mathematically incorrect means to make excellent decisions that bring them into a pure state before each hit.

Later, we are going to invoke a third, more abstract, argument for the fact that people usually make good decisions in the sense of optimal mixed strategy. This argument will follow from the topic of Chapter 13. For the time being, suffice it to say that a group of several persons who make irrational decisions separately may be able to produce a very rational *collective* decision that conforms to the optimal mixed strategy.

Fortuities of Consciousness

The line we have been taking suggests that when our petal-plucking shepherdess cheats, her judge will be lenient. But this is far from certain. Sometimes the judge turns nasty, disqualifies the cheating player, and announces the negative result.

Whether the judge winks at the infraction of the rules or retaliates against it, plucking the petals off the daisy has achieved its aim. The act of conscious observation has taken place, and the loving (or unloving) psyche that had been in a troubled mixed state has turned, at least momentarily, into a pure state. Then everything can start over again, for a mixed strategy is the psyche's (as it is the electron's) natural state.

But if the mixed state is the human psyche's natural mode of existence, how can it make so many simple decisions so easily, without continual worry? Why do we not find ourselves more often in the plight of Buridan's ass, that unfortunate donkey invoked by the fourteenth-century logician Jean Buridan that starved to death between two haystacks because it had no logical basis on which to prefer one to the other?

As opposed to Newton's mechanics, the laws of quantum mechanics do not preclude—theoretically—the possibility that the kitchen table rises from the floor without the application of an external force. Each elementary particle comprising the table moves randomly, and thus it may happen theoretically that each particle finds itself suddenly at a higher position, with the result that the table rises. By the same token, it is also possible that the table suddenly turns into a beagle or into Albert Einstein, provided that there are enough elementary particles available. Both the table and the beagle contain exactly the same elementary particles—what is there to prevent the great metamorphosis?

The great metamorphosis is prevented by the laws of statistics. To be more precise, theoretically, these laws do not prevent this, they only make it *highly improbable*. The probability of the table suddenly rising up by itself is much less than the probability that a monkey seated before a word processor will write the *Odyssey*. A more abstract example that better illustrates the essence of the situation is this: The probability of obtaining fewer than 400,000 or more than 600,000 "heads" when tossing a coin one million times is almost infinitesimally small. The outcome of each toss of the coin is random, yet the combined result of a large number of tosses is highly stable, with the number of heads always hovering around fifty percent.

The position of each electron of a table is random, but their *joint* distribution in space shows great stability, and this joint stability also follows from the laws of quantum mechanics, once the table is there, and thus macroscopic physical laws can be deduced from the ideas of quantum mechanics—at least in principle. Practical calculation would require processing a hopelessly large quantity of data even in the case of a dust particle, but there is no conceptual difficulty. The deterministic laws of the macroscopic world, by which we can calculate the trajectory of a cannonball or the operation of a lathe, do not contradict the principles of quantum physics. But it is not worth applying the calculations of quantum physics to the objects of the macroscopic world, because the calculations cannot be carried out in practice. We could even say that it is a theoretical fact that they cannot be carried out in practice. Nature does not calculate them either; it only makes them work according to its laws.

The situation is similar with our own thinking. Our everyday decisions work as macroscopic phenomena, and thus we can make correct decisions stably, even if the natural mode of existence of our consciousness is a mixed state. But in important questions, when our psyche is deeply affected, our psyche's nature can manifest itself, namely, that it operates according to mixed strategies.

An experienced novelist can write down one sentence after another without long speculation. He can decide relatively easily what the structure of the next sentence should be. Yet many possible twists of the plot are present in his mind simultaneously. The fate of the main characters is in a sort of mixed state until the novel is complete. The process of writing brings the state of the characters into a pure state, and this also affects the psyche of the novelist. There was a period during which the French novelist Gustave Flaubert was visibly unhappy, and when his friends asked what was worrying him, he replied, "Madame Bovary is dead."

The image of the psyche that emerges from these considerations is undoubtedly strange, and the analogy with quantum mechanics seems to have fallen apart. According to quantum mechanics, the behavior of the very small elementary particles, or of systems consisting of a few elementary particles, is strongly nondeterministic, and the closer we get to the macroscopic world, the more often we meet deterministic laws. In the case of the psyche, it is just the other way around. Simple, everyday decisions seem to be more or less determined, while with important questions that affect the whole psyche, mixed strategies that embody the nondeterministic laws become more manifest.

We shall postpone the resolution of this paradox until Chapter 15. For the time being, let us consider the psyche as an entity that behaves according to the ways of quantum mechanics. Physicists were led to problems of consciousness by the internal logic of quantum physics. We were led to similar questions by the logic of game theory applied to human psychology. Game theory is not a new, now only intuited, physics of consciousness; it might be a first step in that direction, though such a new physics might eventually be based on totally different principles. It is also possible that if a new physics of consciousness based on radically new, as yet unsuspected principles is to emerge, we shall nonetheless continue to use game theory to analyze practical decisions, just as we continue to employ Newtonian mechanics to investigate the macroscopic world.

The Methods of Rational Decisions

According to game theory, the correct method of making decisions is as follows. First, determine all the pure strategies for all the participants in the game. Then determine the (objective or subjective) values of all of the possible outcomes; that is, prepare the game table.

Then determine the optimal mixed strategy. When all this is done, make your move by relying on a random-number generator that selects each pure strategy with its assigned probability.

We practically never apply this theoretically faultless procedure in our everyday decisions. Each step of the method encounters severe practical problems. Although we might know the possible pure strategies, it usually happens that new possible strategies occur to us as we ponder a problem. Or we might think of another possible move by our opponent that we had not thought of before, and now we have to include it in our calculations. The subjective values of certain possible outcomes of the game may vary from moment to moment, depending on shifts in our outlook. The calculation of the optimal mixed strategy is made practically impossible by our limited mathematical abilities, either because we do not know the necessary concepts or because the number and difficulty of the calculations is far above our poor power to add or subtract. Finally, in many cases, even if we have been able to calculate precisely the optimal mixed strategy, we may in the end find ourselves unwilling to make an important decision based on a roll of the dice.

Nevertheless, game theory has become an important science (in economic decision-making, for example), because on the one hand it has clarified the elementary steps necessary for rational decision-making and on the other has provided useful methods for carrying out these steps. On the basis of game theory, computer programs have been written that are able not only to perform complex calculations but to assist the decision-maker to restructure the problem so that it can be treated by the methods of game theory. These programs are already widespread because they help in making decisions in problems so complex that they are beyond human capacity to comprehend fully, decisions on which large amounts of money are riding.

Nevertheless, the responsibility for the decision is not taken by the program, nor by its creators, because there is no guarantee that the decision-maker adequately appraised all possible moves and that he correctly determined the values of the possible outcomes. Decision-supporting systems give some help in these processes, but they do not protect against mistakes, nor against the accidental omission of important data. Furthermore, when the decision-maker feeds the various pieces of data into the computer, he cannot tell how they will influence the final outcome. Thus a small amount of uncertain or in-accurate data may have an inordinately large effect on the decision. Although the decision-maker can test many possibilities by starting out from several sets of initial data, and although the program might direct attention to the data whose uncertainty has the greatest effect on the outcome, the ultimate responsibility for making the next move always rests with the person making the decision.

This decision made on the basis of the output of a computer pro-gram is just as much in accord with the decision-maker's under-standing and intuition as is that of the shepherdess deflowering a daisy or of Rabelais's innkeeper. The program can do no more than turn a bundle of hopelessly complex interrelationships into a more concise form that the decision-maker intuitively finds easier to handle.

Creating a decision-supporting system may require several dozen person–years of work. The operation of these systems may be so complex that even their designers may not comprehend them fully, not to mention their users. In a very real sense they are used as ora-cles, which in the case of a given constellation of environment and chosen goals can report the optimum infallibly. Then with the knowledge of these often contradictory optimums we make our still very subjective decisions.

Meditative Techniques

When we wish to make a decision that would promote peace of mind, we are usually unable to verbalize completely just what it is we desire and by what means that desire might be attained. And we are even less capable of verbalizing our deepest feelings, desires, and hopes. This is what makes us so restless. Economic decision-makers have a somewhat easier time of it. At least they have a few clear-cut data points and some well-defined concepts, such as marginal cost of production, exploitation of capacity, excess demand, solvency of cus-tomers, and expected profit. Nonetheless, in attempting to make a rational—or at least a quasi-rational—decision based on computer calculations, this decision-maker would need to use a program the details of whose operation would remain forever a closed book.

Technology helps in the decision-making process, but it cannot replace the decision-maker.

There are a number of psychological techniques that operate in a strikingly similar manner. The procedures of Rabelais's innkeeper and counting off flower petals belong to this group, but there are also more effective and refined methods, including *meditative techniques.* It is a common feature of these methods that they facilitate the natural functioning of the psyche's unconscious internal forces and that they offer an unambiguous reply to our question. Then it is up to us to use this answer as we wish, and this is what makes the decision-supporting computer program analogous to the meditative technique: The economic decision-maker is free to accept the program's output or to run it again with different parameters or to take into consideration other data from other sources. The program's results are significant even if the decision-maker ultimately rejects the advice proffered, since he cannot fully disregard knowing it.

Our psyche is a more refined mechanism than even the most complex decision-supporting program. It has countless capacities, from determining how and when to hit a tennis ball to finding the appropriate words to provide comfort, happiness, or absolution. We do not know how these capabilities work, but they are certainly based on much more general principles than what would be required for solving any one specific problem. How can we use these general abilities to help us solve particular problems?

It is a common feature of most meditative techniques (Buddhist meditation, relaxation, mind control, hypnosis, self-hypnosis, transcendental meditation, etc.) that the meditator focuses attention on an object that is in itself essentially meaningless—perhaps a spot on the wall or our own fingertip.

Hypnosis differs from the other meditative techniques in this incomplete list in that it is the only one that requires the presence of another person, namely, the hypnotist. Nevertheless, there is essentially no psychological or physiological difference between the state brought about by hypnosis and the other meditative states. The common features, however, are abundant, and therefore hypnosis is usually listed among the meditatively induced trance states.

Meditation is usually associated with silence and peace, but they are not its essential condition. For example, the laboratory of Franz

Anton Mesmer (1734–1815)—the Austrian physician who was a pioneer in the development of techniques of hypnosis and whose name has given us the word *mesmerize*—was situated above a smithy, and Mesmer was able to incorporate the noise from the workshop into his technique of inducing a hypnotic state.

Hypnosis is perhaps the most appropriate of the meditative techniques for studying the induced states of consciousness, because experimental conditions can to some extent be controlled and reproduced. Research on hypnosis has shown that most of the conditions generally thought necessary for meditation can be omitted in inducing a trance state, including physical and mental relaxation.

In experiments conducted by Éva I. Bányai and Ernest R. Hilgard, the subjects to be hypnotized were seated on a bicycle ergometer, and while the subject pedaled the bicycle more and more intensively, the hypnotist repeated just the opposite of the usual text of traditional hypnosis. For example, instead of saying, "Your eyes are tired. The heaviness in your eyelids is increasing. Soon you will not be able to keep your eyes open. Soon your eyes will close of themselves. Your eyelids will be too heavy to keep open. Your eyes are tired," they said, "Pedaling has made your legs move quite automatically, without effort. Your legs move more and more easily. Soon you will no longer be able to stop them. Soon they will move by themselves. Your legs will keep pedaling and not be able to stop. Your legs are now even fresher from pedaling."

The hypnotic state was induced in this way in about the same persons and to about the same extent as with traditional hypnosis.

About the same, but not completely. Active-alert hypnosis on the bicycle ergometer succeeded in bringing some persons into hypnosis for whom traditional hypnosis did not work well. The reverse was also true: Some subjects were less susceptible to this method for whom the traditional form of hypnosis was effective. According to the "grandmasters" of hypnosis, everyone is capable of entering this trance state; only a suitable technique has to be found. The traditional methods as well as the active-alert technique work well for a large proportion of subjects, but some individuals are susceptible only to other methods. There are those who enter this state at the moment they lose themselves in contemplation of a scientific problem. We do not understand the ways of meditation. It is a common theme in almost every technique that attention is extremely focused and concentrated.

The Scientific Foundations of Meditative Techniques

Since the so-called mind-control techniques became fashionable it has become generally accepted that meditative states of consciousness are linked to certain electric waves of the brain (the so-called alpha waves). These waves are collateral to meditative states in most cases, and experimenters have shown that trance states can develop without them. Although the theory that linked alpha waves to mind-control techniques has proved false, those methods that have the subject focus on alpha waves work excellently for many people, perhaps because the prestige of science is such that we are inclined to give credence to the scientific notion of alpha waves even if we have no idea what they are. Ultimately, it does not matter whether we concentrate on God, the Universe, the alpha waves, or our fingertips. The important thing is to focus our attention completely on something that is utterly unrelated to the problem that currently occupies our psyche. There are many paths to meditation. Plucking petals and Rabelaisian dice-rolling can also be considered meditative techniques.

Many people consider mind control and other popular "quick and easy" methods as "working" for them. It is not necessary that this effectiveness be placed on a firm scientific footing. For an engineer, for instance, whether something works "in theory" is of secondary importance. Theory is useful if it produces useful technology and if it directs creative thinking. Familiarity with scientific theories can help the engineer to avoid unproductive paths. Yet ultimately, the question is whether a device does or does not work. If it works, it doesn't matter whether the scientific basis is unclear. Even a false scientific theory may lead to revolutionary discoveries. In the seventeenth century, for instance, the fundamentally false theory of phlogiston led to new developments in metallurgy. Similarly, an incorrect scientific theory may provide the basis for an effective meditative technique, especially if the only currently available option is no scientific theory at all.

Some people are able to develop their own meditative techniques without specific previous knowledge and can apply them to difficult life situations. Just as there are persons who have a talent for mathematics—those for whom without special tutelage or even despite the calculation-centered education in schools mathematical structures become alive and meaningful—there are those who have a talent for meditation. These people are able to develop their own effective personalized meditative techniques already in childhood. Unfortunately, in Western culture the pathways open to the realization of such talent are much less well paved than the road to fulfilling one's mathematical, artistic, or athletic talent.

Ideomotor Techniques

Before applying an ideomotor technique (one that creates an unconscious or involuntary bodily movement in response to a thought or an idea rather than to a sensory stimulus), the meditating person usually decides in advance that when the answer to the question posed settles in his psyche, the psyche will indicate this by some automatic movement. For instance, depending on whether the answer is affirmative or negative, the index finger of the left or right hand, respectively, will move of its own accord. *Ideomotor* means that movement becomes one with the thought, as, for example, when you steer a car at high speed, since if you manipulate the steering wheel with too much conscious intervention, you will almost certainly get into trouble. It is sufficient just to think in the rightward direction and the car will move to the right as if by itself—due to ideomotor steering.

Ideomotor techniques are in harmony with the hypothesis, verified by a number of experiments, that thinking in not primarily verbal. In deep states of meditation it may be quite difficult to utter as simple a word as *yes* or *no*. Furthermore, if the answer to a question really comes from a mixed state of the psyche, then the full answer would not be a simple *yes* or *no*, but a certain probability distribution. Thus, the truth would actually be masked by a simple *yes* or *no*. The situation is much more complex, and only the full description of a mixed state would reflect the real truth. Such a description, however, would make the decision itself, the selection of a pure strategy, difficult.

Another well-known application of the ideomotor technique is to take as the alternatives the right and left sides of the body. In this case, the ideomotor response may appear as a slight (but sometimes a spectacular) leaning toward one side. It is an advantage of this technique that it is not essential that the subject imagine the alternatives directly, for they can be projected onto a screen, which promotes the necessary distancing.

The use of a pendulum is often a highly successful technique in obtaining an ideomotor response. The interpretation of the pendulum's movements are usually determined in advance. For example, motion toward and away from the body will mean *yes*, while right-to-left movement means *no*. It is an advantage of the pendulum that it magnifies even very small ideomotor movements so much that they become directly observable. This method is also used when we do not want to get an answer directly to a specific question but want rather to understand more clearly our feelings and motivations. In these cases two additional responses are allowed: The pendulum circulating clockwise means "I don't know," while the counterclockwise direction means "I do not wish to reply."

The pendulum method has proved to be an excellent technique of meditative cognition, although it does not follow from any known scientific law. Perhaps we shall discover that the origin of the legends of magic pendulums and magic wands lies in the fact that some pendulum devotees are excellent talents at meditation, who are able to consider very small signs in the environment—undetectable by others—when making their unconscious decisions.

The pendulum is also suitable for distancing. The user of the pendulum does not even have to be aware that it is actually he who has made the decision and not the pendulum. The users of "magic pendulums" generally create complex, confused, and clearly incorrect "physical" theories about what actually moves the pendulum. Nevertheless, the pendulum may produce "miracles" in the hand of a very talented person, just as a paintbrush does in the hand of a talented artist.

Today Thus, Tomorrow Otherwise

By means of meditative techniques a more or less definite answer to our question, that is, a pure strategy, can be obtained from the psyche, which is by nature in a mixed state. To some extent, we may believe that this answer is given by the psyche on the basis of its best, most subtly developed, abilities. But since the reply comes from a mixed strategy, it can be different on repeated posing of the same question, just as electrons fired under the same circumstances may be found at different places. Therefore, the masters of meditative techniques usually advise not to ask the same question repeatedly within a short interval of time. The profoundness of this empirical suggestion can be explained by game theory. Once we know the optimal mixed strategy and have made our decision based on it, repeatedly rolling the dice (instead of accepting the first outcome) will distort the optimal probabilities themselves, thus making the whole procedure pointless.

This does not contradict the observation that the psyche often "rolls the dice" in many steps or even "cheats" with repeated rolls just by changing our moods, or perhaps by the aid of meditation. We use what we have. If, for example, we have a pair of six-sided dice and we would like to say *yes* or *no* with respective probabilities 1/36 and 35/36, we can calmly roll the dice and say *yes* only if we throw a double six, arriving at exactly the same outcome as if we had thrown a die of 36 sides.

Meditative cognition does not give a correct solution in a single situation; it does so only in the long run, when the *totality* of our decisions, perhaps our whole way of living, is considered. This is no wonder, since the situation is the same in purely rational game theory. If in some game the optimal strategy is a mixed strategy, then there exists no single correct decision, and the correctness of a given strategy can be studied only over the long run. It is possible to win in any game with the wrong tactics, but in the long run, however, only good tactics will prove successful.

What we have said about meditative techniques is true of healthy individuals, free of severe psychological problems. Psychologists and psychiatrists also apply such techniques in order to uncover latent factors (such as the unconscious effect of a forgotten childhood trauma) that pathologically disturb the normally mixed state of the psyche and to help bring the psyche into equilibrium

Conversely, one has to tread very carefully in the psychoanalysis of healthy individuals. If a person has no disturbing condition that would prevent the normal (that is, based on mixed strategy) conduct of his life, psychoanalysis might disrupt that normality, because it could bring to the surface factors that have been built into his mechanisms of decision-making, interfering with more or less well functioning mixed strategies, thereby destroying a carefully balanced equilibrium.

Logic and Intuition

If a problem can be treated within the framework of pure rationality, it is not worth looking for another solution, for with pure reason we can learn not only the *outcome*, but also exactly why what we know is true. Our results can then be precisely explained to others, and since we also know the logical process leading to the given result, others can follow it on the basis of our report. This is why logical, rational reasoning is so convincing.

It follows from Gödel's theorem that the rational approach cannot work for every problem, no matter what specific concept of rationality we use. But we don't have to wait for a Gödelian problem before encountering practical difficulties. Our concepts about how nature or our psyche works are incidental human concepts; even stable concepts like place and speed may not express entities that really exist in nature. The more thoroughly we get to know the operation and correct usage of a concept, the more nonplused we become if the concept somehow suddenly betrays us, if all of a sudden it cannot satisfactorily explain even the most simple phenomenon. Yet such occurs not only in quantum mechanics; even our basic mathematical concepts may bolt. Let us look at an example.

Imagine that Rabelais's innkeeper cheats, but not by rolling the dice in a duplicitous way, but by rolling the dice repeatedly until the result he likes dominates. Suppose, for example, that he wants the dice to show an even number, but he rolls an odd number. He now continues to roll until more evens than odds have been thrown. If he rolls the dice with infinite patience, what are the chances that he can eventually stop rolling the dice? We first calculate the separate probabilities for when "evens" will first dominate after the third roll, after the fifth roll, after the seventh roll, etc. (Even numbers cannot dominate *for the first time* after an even number of rolls, since an odd number was thrown first.) We shall not go into the details of the calculation, but it turns out that the sum of this infinite number of probabilities is 1, which means that an even number will dominate sooner or later with a probability 100%.

If the even numbers will be dominant sooner or later with probability 100%, after how many rolls is this expected to happen? How many times will Rabelais's innkeeper have to throw the dice on average if he first throws an odd number? If our innkeeper throws two even numbers immediately after the first odd number, he will have finished after the third round, but this happens only with probability 1/4. If we now add up the infinite number of expectations (we shall again skip the math), then to our greatest surprise this sum is infinite.

It follows from our logically perfect system that the prevalence of even numbers is *certain to take place*, but this event is expected to take place *after an infinitely long period*, that is, in all likelihood never.

My brain understands this calculation perfectly well, and I also know that the system in which we made the calculation is logically faultless, but I do not understand the result. And even though I *know* that the outcome is true, I do not understand how the world can be like this, a world in which such weirdness may occur.

I asked a few of my mathematician friends who did not know this exercise to estimate the result without specific calculations, just relying on their mathematical intuition. The majority predicted the correct answer to the two questions, but none of them could explain it in common human concepts. When they were governed by their mathematical intuition, they arrived at the correct result, but their *human* intuition did not understand it either: How can something be expected to take place with certainty yet on average never occur? We trust our lives to the results of similar calculations—we peacefully board an airplane or turn on the lights—without being able to answer this simple question.

Cognition based on meditation offers just the opposite experiences. Our psyche often gives definite, clear-cut, and useful answers that our brain does not really understand and that we cannot explain rationally. Logic and intuition work in us in complementary ways, but neither of them follows from the other. In fact, most of the time they do not even understand each other. But taken as a whole, this entire psychic system may be completely rational, or at least quasi-rational.

After the previous mathematical example it becomes clear that the epigraph to this chapter (*My brain understands it, but I don't*) expresses a real phenomenon. Yet it also became evident from our analysis of meditative techniques that the opposite (I understand it, but my brain doesn't) is also a real phenomenon. We may recall a formulation by Niels Bohr: "The opposite of a correct statement is a false statement. But the opposite of a profound truth may well be another profound truth." Logic and intuition work in our mind as antitheses, as profound truths.

For rational thinking based on pure, logical reason, everything that is not completely rational is considered irrational. But in a deep meditative state, it is logic, pure rationality, that is considered irrational, as leading nowhere. If we do not want to talk about rationality *itself*, but about the *psychology* of rationality, we cannot avoid talking about the *sensible* forms of irrationality, since they are psychological phenomena that are not irrational at all.

13

Collective Rationality

They look for you in the best hiding place first.

In 1984, the editors of the journal *Science 84* planned to announce a game in which every reader could apply for either $20 or $100. If fewer than 20% of the readers applied for $100, then everybody would receive the money applied for. If, however, more than 20% of the readers applied for $100, then nobody would receive anything. Unfortunately, the owner of the journal was afraid of losing his shirt, and the game was not announced. It seems that Lloyds of London had declined to insure the journal against loss: "Sorry and all that, you know. Just too great a degree of risk." In the end, the editors explained their idea and why they couldn't proceed as planned. They announced the same game but in a hypothetical form. They invited readers to write in how much they would have applied for if there had been a real competition.

Although there was nothing to be won, more than thirty thousand "applications" arrived: 65% wrote $20, 35% wrote $100. Thus *Science 84* would not have had to pay anything. Lloyds appears to have been somewhat overcautious.

Isaac Asimov, the famous science-fiction writer, wrote a letter to the editors before the results had been published: "A reader is asked to check off $20 and consider himself a 'nice guy,' or check off $100 and consider himself *not* a nice guy. In such a case, everyone is going to take that option of pinning the nice guy label on himself, *since it costs no money to do so*." Asimov probably thought this way because the competition was announced following an article about cooperation. Well, Asimov was wrong.

Since the results are known, perhaps the journal would not after all have been taking much of a risk in a game played for real money, since in this case as well probably at least 35% of the participants would have applied for $100. In fact, if Asimov's supposition has even a kernel of truth in it, then we can expect that even more than 35% would have asked for the larger sum.

Looking at it from another perspective, however, the risk is really great in such a situation. The greater the number of competitors, the more the journal stands to lose. A greedy millionaire might send millions of applications for $20, which would almost guarantee that at least 80% of the applications were for $20. Thus, by spending a million dollars or so on postage, he could win a few million times $20. The editors may prevent this by including the proviso that an application is accepted only if accompanied by a coupon clipped from the journal. This would limit the number of applicants to the number of copies sold. Yet a well-organized syndicate with solid capital might buy up every copy of the journal for $4.95 a pop and then send 20% of the applications for $100 and the remaining 80% for $20. But this syndicate could turn a tidy profit even if it were very cautious, indeed positively cooperative, and sent in only $20 applications. This way, even if the few readers able to obtain a copy applied for $100, the syndicate would pocket a few million dollars.

The readers might also cooperate by organizing themselves into groups of five, agreeing that four of them would apply for $20 and the fifth for $100. Then they would divide the spoils, giving everyone a $36 profit. Although theoretically this is possible, practically, this can be considered an unfeasible strategy. Who would organize such a cooperative effort? Most readers would probably not begin to get organized but simply ponder over whether or not to enter the competition, and if yes, for how much money. The game would take its course, and the journal would have given its readers (without considerable financial loss) a little mental stimulation, which to the journal would have been worth the insurance premium.

Analysis of the *Science 84* Game

In many respects, the game of *Science 84* resembles *Scientific American*'s million dollar game. They are both mixed-motivation games, since on the one hand, it is the interest of the players as a community for no more than 20% of them to apply for $100; otherwise, nobody wins anything. But on the other hand, it is everyone's individual interest to be among the $100 applicants if everybody wins.

If fewer than 20% of the applicants applied for $100, then every individual who had observed the interest of the common good would regret having not been more assertive. However, if 25%, say, of the readers applied for $100, then if I, say, had been one of these noncooperative applicants, then I personally need feel no remorse, since even if I had been more modest, nobody would have won anything anyway. Thus, nobody can be blamed for the fact that the readership of *Science 84* failed to win several million dollars, since the responsibility is divided among the many assertive players, and almost no onus falls on any single player.

This game is a dilemma, too, like the million dollar game and the problem of common pastures that we saw in Chapter 3. The structure of the game of *Science 84*, however, is different from the previous ones. It is a multiperson version of chicken. For me, as an individual, the most favorable case is if I compete and apply for $100, while at least 80% of the others cooperate. The second-best case is if I along with 80% of the others cooperate, because I at least win $20. The worst case is if I compete and so do at least 20% of the others, because nobody will win anything this way. It is only slightly better if I cooperate and more than 20% of the others compete, because although nobody wins anything, I can consider myself a moral winner. All the other players arrive at this same conclusion. The hierarchy of values is identical with that of chicken.

This also shows that the game of *Science 84* is a trap to its core. In cases like this, the categorical imperative generally requires ethical behavior to realize a mixed strategy. What could this mixed strategy be?

As we saw at the end of the previous chapter, our mathematical intuition can be fallible. The game of *Science 84* also proved this. Most people, having acquainted themselves with the concept of mixed strategies, would determine that the best course of action is to compete with a probability of 20% for the $100 (for example, by throwing a five-sided die and applying for $100 only if the die shows 5). Yet this is far from the optimal strategy.

If everybody applies for the $100 with a probability of 20%, the chances of nobody winning anything is about 50%, since the probabilities of slightly more and slightly less than 20% of the people applying for the $100 are equally likely. If the players restrain themselves just a little, the chances of winning radically increase. If there are 10,000 applicants, the optimal strategy is to apply for the $100 with a probability of about 18%. In this case the probability of nobody winning will be much less than 1%, while the chances that any particular individual gets to apply for $100 decrease only a little.

The result is even more surprising if there are only 5 players, in which case the optimal mixed strategy is for everyone to apply for the $100 with a probability of only 10%! If everybody applied for the $100 with a probability of 20% (as our intuition suggested), then in half of the games nobody would win anything, and the expected gain of every individual would be $18 in the long run, which is less than the $20 to be gained by complete cooperation. However, if everybody applied for the $100 with a probability of only 10%, then in the long run, the average profit of every individual would rise to above $25 per game!

Before continuing our discussion of the game of *Science 84*, let me settle an old debt from Chapter 2: Let's see what actually happened in the million dollar game of *Scientific American*.

Outcome of the Million Dollar Game

It is likely that the editors at *Scientific American* considered the million dollar game risky, and so they announced the competition in a slightly different form from that described in Chapter 2. There was a small, but significant, difference: An entrant was allowed to compete as many times as desired by making a single application. If, for instance, someone wanted to compete one thousand times, his chances of winning increased a thousandfold, but then, of course, the maximum of his potential winnings dropped to $1,000 (if nobody else entered!). For the publishers this made the game much less of a risk: Certainly, at least one player (the publisher himself *in extremis?*) would manage to spoil the game for all. And that is just what happened. Someone spoiled the game. More than one.

More than two thousand applications arrived, with the following multiplicities:

number of times they entered	number of applicants
1	1133
2	31
3	16
4	8
5	16
6	0
7	9
8	1
9	1
10	49
100	61
1000	46

In addition, thirty-three people entered one million times and 11 one billion times. Nine players entered a googol (10^{100}) times, and fourteen a googolplex ($10^{10^{100}}$). There were similar astronomical figures as well.

Those who wrote such large numerals can safely be called spoilers of the game. They fall totally outside the purview of game theory, although psychologists may perhaps be interested in them from another aspect. The aim of many spoilers was not actually to spoil the game, but to display their knowledge to this authoritative journal. Some entrants described gigantic numbers by complex definitions, such as numbers arising in the mathematical proof of Gödel's theorem, or Avogadro's number. Still others simply wanted to be declared the winner, even if the prize was $0.00. The greater the number of entries, the greater the chances of winning. *Scientific American*, however, was not amused, and no drawing was made, on the grounds that there was nothing to give to a potential winner.

The Dutch cultural historian Johan Huizinga wrote in his book *Homo ludens:*

> The game spoiler is something quite different from the cheater, who acts as if he were really playing the game and never ceases to recognize the enchanted circle of the game. Him the other players forgive more readily than the game spoiler, for the latter tramples upon the game's very world. . . . Therefore, he must be destroyed, for he threatens the existence of those who play the game. . . . Even in the world of serious purpose the cheaters, hypocrites, and tricksters have an easier time of it than the game spoilers: the apostates, the heretics and innovators, and the prisoners of conscience.

The spoilers of *Scientific American*'s game had no fear that the other entrants would take their revenge. They could act anonymously, without risk, and perhaps this is why their proportion was relatively large. But what is the moral status of those who entered only a couple of times, perhaps only a modest once? They, too, are condemned by Douglas R. Hofstadter, the spiritual father of the game:

> Curiously, many if not most of the people who submitted just one entry patted themselves on the back for being "cooperators." Hogwash! The *real* cooperators were among those 10,000 or so avid readers who calculated the proper number of faces of the die, used a random-number table or something equivalent, and then—most likely—rolled themselves out in this way. I appreciated hearing from them. It is conceivable, just barely, that among the thousand-plus entries of "1" there was one that came from a super-rational cooperator—but I doubt it. The people who simply withdrew *without* throwing a die I would characterize as well-meaning, but a bit lazy, not true cooperators—something like people who simply contribute money to a political cause but then don't want to be bothered any longer about it. It's the lazy way of claiming cooperation.

We, however, state just the opposite of Hofstadter's last assertion. Neither nature nor human reason realizes chance by rolling real dice, *even if they succeed in achieving an optimal mixed strategy.* Nature achieves its ends by quantum-physical, genetic, and other mechanisms, including what it has developed in our thinking: that we make our decisions on the basis of our capricious moods and feelings, perhaps without attentive reasoning, relying on not fully rational means of thinking, such as meditative techniques. We rely on methods that Hofstadter calls "lazy." And yet, if these methods help us realize an approximately optimal mixed strategy, we should call them economical, effective, and quick. But this is not what at first glance the actual results seem to indicate.

The Hidden Aim of the Game

That there were so many spoilers of the million dollar game was most likely caused by the way the competition was announced. Had *Scientific American* announced the competition in its original form, the spoilers would have had the rug pulled out from under them. It would not have been worth spending millions of dollars on stamps, never mind trying to locate a googolplex stamps and envelopes on short notice. In this case, the game would have cost *Scientific American* only a few hundred dollars, but such an outcome could not have been foreseen. Anyway, the significance of the game as it was actually played as a psychological experiment is rather limited, for it combined two completely different games. The goal of one was to spoil the game, while that of the other was the usual sort of game. Nevertheless, let us disregard spoilers of the game and also that the possibility that the game could be spoiled may have affected the behavior of other players, and let us just examine to what extent the behavior of the totality of the other players can be considered rational or irrational.

Even if we disregard the few hundred spoilers of the game, the number of applicants who applied only once or only a few times, about twelve hundred, is still far from being optimal. The purely rational considerations in Chapter 2 demonstrated that *Scientific American* should have expected to lose more than half a million dollars if the optimal mixed strategy were followed, and the chances of their getting away with a low payoff were slim. That there were more than eleven hundred single applications proves that the totality of readers did not come close to realizing this optimal mixed strategy.

On the other hand, if there had been only these twelve hundred applicants, the joy of the winner would by no means have been spoiled completely. The amount of this joy would have been about the same as in more modest competitions. There is nothing wrong with winning a few hundred dollars. After all, winning a million dollars belongs more to the realm of dreams. If we want to think well of the totality of the players, discounting the spoilers on account of the way in which the competition was announced, we may conclude that the players did not spoil the game *completely*; they indeed showed some collective rationality. So perhaps *Scientific American* should not have offered a million dollars. A thousand would have been fine. In that case we could conclude that the players together approximated the optimal mixed strategy quite well.

Although we have no experimental evidence that a smaller prize would have served as well, it has been proven by many similar psychological experiments that such an attitude is often valid. The psychological effect of a stimulus is not directly proportional to its magnitude. Weber and Fechner's almost 150-year-old law regarding the sensation of most stimuli—whose magnitude we sense by comparing their effects to some already existing state—is still valid. This law states that the subjective sensation elicited by a stimulus is *logarithmically* proportional to the physical magnitude of the stimulus. Thus, if doubling the stimulus increases the sensation by one unit, then quadrupling the stimulus increases the sensation by only two units. This law holds true of the subjective sensation of relative weights and light intensities just as much as of money rewards, of fines, or even of the effects of doses of medicine. For in these cases, we unintentionally make comparisons with an already existing state, such as our present financial state or current physical condition. When economists hypothesized the existence of the law of diminishing returns in the Arrow–Debreu equilibrium model (mentioned in Chapter 9), they were happy to hit upon the Weber–Fechner law in justifying their assumptions.

According to this law, the difference between the subjective values of winning one thousand dollars and one million dollars may be much less than a thousandfold. The difference is evidently significant, and thus our previous assumption about the rational behavior of the players in the million dollar game was too generous, but it was not completely unrealistic. In the case of the million dollar game, the true, subjective, hidden aim of the game might have been more than just winning itself. It might well have been the pleasure of winning plus the receipt of a few hundred dollars. If this is true, then as a group the players exhibited fairly rational behavior.

The Hiding Lottery

In order to validate this line of reasoning, it would be good to find evidence that was less speculative and more direct. For this purpose I designed a game called the *hiding lottery*.

In this game, a lottery ticket has to be filled in as follows. One marks six numbers in the range 1 to 49 that are displayed in a 7 × 7 grid. The winner is not drawn as in a usual lottery but is the player whose numbers resemble least those of the other players. To quantify this notion, in the case of each player we add up how many other players marked the same numbers as he did. The player who achieves the smallest number is the winner.

We tested this game many times in groups of 10–30 players, and we also announced it in the Hungarian magazine *Élet és Tudomány* (*Life and Science*). We received 236 entries. In Table 1 is listed the distribution of the numbers chosen.

Table 1

1	2	3	4	5	6	7
39	**40**	**30**	**29**	**30**	**33**	**36**
8	9	10	11	12	13	14
28	**14**	**26**	**31**	**19**	**35**	**26**
15	16	17	18	19	20	21
27	**22**	**45**	**26**	**36**	**30**	**29**
22	23	24	25	26	27	28
23	**36**	**24**	**28**	**34**	**28**	**32**
29	30	31	32	33	34	35
30	**24**	**40**	**31**	**20**	**21**	**17**
36	37	38	39	40	41	42
17	**31**	**21**	**27**	**26**	**26**	**25**
43	44	45	46	47	48	49
39	**28**	**23**	**18**	**42**	**37**	**37**

The frequencies of the individual numbers are not quite random, but they are close to it. For the sake of comparison, we also show in Table 2 the frequency of numbers drawn in the Hungarian lottery (5 out of 90 numbers) from its beginning in 1957 to 1995.

Table 2 illustrates the magnitude of fluctuations blind chance can cause even in such a large number of draws (over two thousand). It can be calculated statistically that the joint result of the players of hiding lottery was not far from each player trusting his selection to chance.

Table 2

1	2	3	4	5	6	7	8	9	10	11	12	13	14	15
117	96	142	112	100	118	129	117	110	138	108	130	136	114	127
16	17	18	19	20	21	22	23	24	25	26	27	28	29	30
114	107	132	122	119	119	123	131	121	120	104	100	100	142	95
31	32	33	34	35	36	37	38	39	40	41	42	43	44	45
97	116	115	130	120	124	117	119	97	97	118	131	118	111	117
46	47	48	49	50	51	52	53	54	55	56	57	58	59	60
123	135	113	133	114	133	109	114	117	108	132	107	101	106	123
61	62	63	64	65	66	67	68	69	70	71	72	73	74	75
110	107	98	137	125	114	120	110	133	100	119	117	117	109	143
76	77	78	79	80	81	82	83	84	85	86	87	88	89	90
117	138	122	115	115	121	105	115	122	108	141	112	94	107	122

But what does this have to do with collective rationality? Assume that the survival of a species depends on some of its individuals winning at hiding lottery occasionally, other species also being participants in this game. If the individuals of a given species consistently did not choose certain numbers, they would offer the possibility to rival species to win more often. A clever rival species would not mark these numbers with high frequency, because this would decrease their own chances. The members of this smarter species would mark every number with equal probability, including those that are less frequently marked by the intellectually more challenged species. These neglected figures would be winning numbers, and since the smarter species chooses these numbers, the winners would more often come from this species, who would thereby gain a survival advantage against their less clever rivals.

A random strategy in such a game is an *evolutionarily stable strategy*, since it is widespread among the individuals of a species, and no other strategy can be more successful than this. Thus, this can be considered the most rational strategy.

The group of 236 players of *Élet és Tudomány* realized this optimal strategy quite well. Yet probably only a few players chose their numbers at random. Most, perhaps even all, of the players chose their numbers as a result of some nonrandom reasoning process. Since the only rational strategy is random selection, this also means that the reasoning of most of the players—whatever it was—cannot be considered rational. Nevertheless, from a higher perspective (namely, that of an evolutionarily stable strategy), the collective outcome can be considered rather rational.

In the groups that played our game we asked several players how they had selected their numbers. None of our subjects trusted their decisions to blind chance. They all applied some kind of logically supported line of reasoning, like trying to avoid the so-called "lucky numbers" or the more conspicuous figures at the edges (or the middle) of the 7×7 grid, on the assumption that they would be more probably chosen by others. Others chose just these numbers, reasoning that the others would neglect these very figures. Quite a few players did not trouble to try to understand the essence of the game and just marked the first numbers that occurred to them or just marked their favorite numbers.

In one version of the game, there were no numbers indicated on the grid, the players having to cross out any six of the 49 empty squares, while in another version the players had to select six numbers between 1 and 49 without being shown a table. In these two versions, the collective outcome was much less random than when both the arithmetical and the geometric aspects were present. The more aspects available to the players for selection, the better they were able to approximate the optimal mixed strategy collectively.

Generally, real-life problems offer a large number of aspects for consideration. Often, it is not clear to people what these aspects are. Thus, there is quite a large scope for many kinds of thinking to develop, and their sum may lead to optimal, evolutionarily stable collective strategies. To study this phenomenon further, it is worth considering a more complex game.

The *Smallest Number Wins* Game

In hiding lottery, the mathematical formula of the evolutionarily stable strategy was exceedingly simple: The probability of marking each number was the same. For instance, we could draw the numbers from a hat. This embarrassingly simple strategy can be optimal, because the rules of the game make no distinction among the numbers. Nevertheless, the players do make distinctions. For instance, they distinguish numbers at the edges of the grid from those in the middle, or they single out lucky numbers. This differentiation, however, does not follow directly from the rules of the game, but from the fact that while the players think about the numbers to be chosen, they make assumptions about the possible strategies of the other players. As we have seen, this procedure is not rational, no matter how logical it may seem, since the only rational strategy is random selection. Yet the *sum* of these nonrational strategies approaches the optimal mixed strategy. Therefore, we may consider the players' reasoning processes as quasi-rational, more or less.

The optimal mixed strategy of the *smallest number wins* game is far from being so simple. This game was announced in *Füles*, a Hungarian puzzle magazine, with the following rules. The participants in the game were to choose a positive integer and write it on the coupon that appeared in the journal. The winner was to be the player who sent in the smallest number that was not sent in by anybody else.

In this game, the various integers already have diverse personalities. The small numbers are attractive, because if we happen to be the only one to send in that particular number, chances are that we will be the winner. Yet large numbers are also not without their charms, as it may easily happen that the smaller numbers are neutralized by many players choosing them, and thus we may win with quite a large number. Perhaps we should choose one million. But if somebody is cleverer than us and chooses 999,999, then he may win instead of us. So perhaps we should outfox them and choose a smaller large number? With such a line of reasoning we might soon find ourselves down among the small integers.

There is no assured winning number in this game, since any number can be a winner, depending on the choices of the others. But perhaps there is an evolutionarily stable strategy in this game—a mixed strategy according to which every number is selected with a certain probability. Will these probabilities be the same for every number?

Assume for a moment that the biological survival of some species depends on winning in these *smaller number wins* games, and that the numbers can be chosen between 1 and 1,000,000. If our species chooses every number equally often, then members of another species who favored smaller numbers would perhaps win more often.

It is a difficult mathematical problem to calculate precisely the evolutionarily stable strategy of the *smallest number wins* game, but a computer simulation can estimate it quite well. The essence of the simulation is that we start from a hypothetical population each of whose members always chooses a particular number, and we assume that at the outset each number choice is equally represented in the population. Then we select a few beings from the population at random (say, as many as the number of players entering the competition of *Füles*) and let them play a round of the game. One organism that always chooses that winning number will be added to the population, since one of this group has demonstrated reproductive fitness. A consequence of this reproduction, naturally, is that the survival chances of the winner and its offspring in the next rounds will have slightly decreased, since there are now more of its type in the population, increasing the odds that its number will be chosen more than once. If this simulated game is run a few million times on the computer, the relative proportions of organisms representing the different numbers approach limiting values—demonstrating an evolutionarily stable strategy.

To what extent did the players of *Füles* approach this evolutionarily stable strategy? How well did the frequency of numbers sent in by the players approximate the proportions obtained in the computer simulation? If the similarity is great, we may conclude that a large number of players who singly do not think rationally collectively produced a very rational strategy.

The *Füles* competition was entered by 8,192 players. These players demonstrated their strategic diversity by choosing over 2,000 different numbers. The winning number was 120. Every smaller number was chosen by at least four players, except for number 94, which was the choice of only two players. The outcome of the game was presented in detail in my book *Ways of Thinking*, where I concentrated on the different strategies. Quite a few modes of thought were presented, their existence being demonstrated by the results of the competition. Now it is the *collective* behavior of the players that is of interest to us.

If we examine the individual numbers that were chosen, we observe results very different from those predicted by the optimal mixed strategy demonstrated in the computer simulation. For instance, about two-thirds of the entries were odd numbers, although in the evolutionarily stable strategy odd and even numbers appeared in about the same proportion. Lucky numbers (such as 7, 13, 17, 21) and 1 also occurred more frequently than justified by the evolutionarily stable strategy. But if we disregard these special numbers and group the rest of the numbers by tens, then the remaining 80% of the players exhibit the evolutionarily stable strategy fairly exactly.

Thus, the players in this game demonstrated irrational ways of thinking such as selecting magic numbers, or figuring, "I'll kill myself if nobody else chooses 1 and I don't either; so I will." On the other hand, it is evident that the collective thinking of about 80% of the players can be regarded as quasi-rational. (As we know, any reasoning process that is not based on random selection cannot be considered *purely rational!*) For instance, the attractions of small numbers and large numbers were quite thoroughly neutralized in the variety of strategies exhibited. This is why they can be considered quasi-rational—in spite of their apparent nonrationality.

From a psychological perspective, one can imagine a line of reasoning that would lead to a result such as, "I should send in a number between 160 and 170," but our present methods of investigation are not so refined, and in analyzing the evolutionarily stable strategy, we consider all of these strategies as a single strategy. It is thus all the more surprising that the majority of the players collectively produced an evolutionarily stable strategy with remarkable precision. From the point of view of an evolutionarily stable strategy, we are able to predict the collective behavior of people in games like smallest number wins or hiding lottery, or, as we saw in Chapter 9, in segments of the economy governed purely by the rules of the free market.

The Means of Collective Rationality

In hiding lottery or smallest number wins, the players compete against one another with no common interest. In the game of *Science 84*, however, the players also compete against one another, yet it is also their common interest that no more than 20% of them apply for $100. The game of *Science 84*, like the million dollar game or the problem of common pastures, is a game with mixed motivations.

It was suspected since Adam Smith, and it has been known since John von Neumann, that a purely selfish strategy may lead to equilibrium in purely competitive games. This was more or less supported by hiding lottery and smallest number wins, with the additional finding that many different quasi-rational ways of thinking can also lead to a purely selfish strategy. Purely rational strategies, based on game theory and rolling the dice, are not necessarily the only practical means of realizing a collective rationality.

The presence of collective rationality was not so clear-cut in the million dollar game, although after discovering the subjective, hidden aim of the game, we no longer have to discard the possibility that collective rationality is operating in this game, too. In the game of *Science 84*, however, it became clear that collective rationality does not always develop spontaneously. The proportion of 35% of players applying for $100 is very far from optimal, and the result was that nobody won anything, although everybody could have won something. The psychological studies of the problem of common pastures have led to similar conclusions. The results of these experiments were essentially the same as those of the dollar auction and of the prisoner's dilemma experiments discussed in Chapter 3. In the majority of experimental groups, every cow eventually starved to death.

As we saw in Chapter 4, two of the four types of games with mixed motivation (prisoner's dilemma and chicken) include a difficult problem that cannot be solved by the means of purely individual rationality. The experimental evidence in the two-person versions of these two games shows no collective rationality. It can be demonstrated that the million dollar game is actually a multiperson version of the gift of the Magi, and individual, quasi-rational strategies in games of this type can achieve quite acceptable results that do not interfere with collective rationality too much.

In the cases of the two difficult trap situations, only general ethical principles that point beyond individual rationality could help, like the golden rule or the categorical imperative. It is possible that there exists an effective ethical principle even more general than the categorical imperative. But it is also possible that no such principle will be necessary, for as we have seen, as opposed to the golden rule, the principle of the categorical imperative can be reconciled with mixed strategies and thus with the unrestricted validity of game theory.

There are two possible roads to the realization of collective rationality. One is pure rationality, the complete integration of general principles (like the categorical imperative) necessary for collective rationality into our system of thinking followed by a perfect application of game theory. The other possibility is the development of appropriate quasi-rational strategies. Human thinking seems to function more according to the latter method—we shall talk about this in greater detail in the next chapter. However, the achievements of pure rationality may also contribute to the development of these quasi-rational strategies, but not by making use of them directly, rather by taking root deep in our thinking; and the purely rational principles—like the conservation of matter, or gravity—become involuntary parts of our everyday, quasi-rational modes of thought. These quasi-rational strategies can appear in the deeper understanding of such concepts as self-control, tolerance, and the consideration of others' points of view.

Experiments with the prisoner's dilemma or the results of the game of *Science 84* have shown that these quasi-rational strategies of ours are far from being perfect vehicles for the realization of collective rationality. Nevertheless, these very same experiments have also shown that these strategies are present to some degree in our ways of thinking. Almost half of the players chose cooperation in the two-person prisoner's dilemma games, and perhaps the 35% who went for the big bucks in the game of *Science 84* is not hopelessly far from the ideal of 20%.

According to Kant, the categorical imperative is inherent in man. (To tell the truth, he thought that this was true of logic, too, but the fact that psychological experiments have proven the latter to be false does not mean that the former is also false.) It is possible that the (at least limited) operation of the categorical imperative developed in man through the group-selection component of evolution. Nevertheless, we had to wait until the 1700s for Kant to put this extremely effective generalization of the golden rule into words. Perhaps game theory had to come along before we recognized the real significance of the categorical imperative, including what we said in Chapter 4: that the categorical imperative is not only more general than the golden rule, but it is essentially different; for instance, it is logically consistent with diversity in biology and in human thought.

These discoveries and concepts are still relatively new. We have yet to integrate them deeply into our everyday, quasi-rational methods of thinking, which could lead to our being able to spare ourselves the increasingly dangerous consequences of pernicious traps, such as hatred, the destruction caused by intolerance, and the irrevocable ruin of the environment.

14

The Heterogeneity of Human Thought

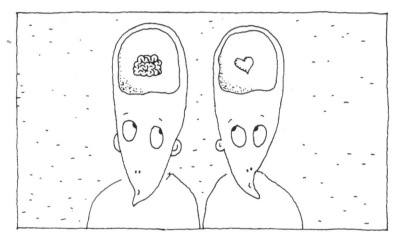

It is rational that human thinking is not rational.

The great Swiss psychiatrist Carl Gustav Jung wrote the following in 1925, three years before the birth of game theory:

> Our will is a function regulated by reflection; hence it is dependent on the quality of that reflection. This, if it really is reflection, is supposed to be rational, i.e., in accord with reason. But has it ever been shown, or will it ever be, that life and fate are in accord with reason, that they too are rational? We have on the contrary good grounds for supposing that they are irrational, or rather that in the last resort they are grounded beyond human reason. The irrationality of events is shown in what we call *chance*, which we are obviously

compelled to deny because we cannot in principle think of any process that is not causal and necessary, whence it follows that it cannot happen by chance. ("Two Essays on Analytical Psychology," p. 49)

Before citing Jung further, let me note two things and then quote the footnote inserted by Jung at this point. First, a new thought—no matter how radically new it appears to be—is usually already "in the air" when it is born. Jung's idea cited here undoubtedly forecasts the spirit of game theory, although in a totally different language and from a remote branch of science.

Second, Jung considered chance as a form of irrationality, which seems to be very logical, since what could be less rational than blind chance. Since then, however, we have learned that chance may be a means of realizing pure rationality; in fact, sometimes this is its *only* means. Thus, the correct use of chance (for example, employing the optimal mixed strategy) can be considered perfectly rational, while its incorrect use (for example, using incorrect probabilities) can be seen as definitely irrational. The approximately correct use of chance is called *quasi-rational*, even if this approximately correct use is not guided by purely rational principles.

Jung did not make a clear distinction between *rational, irrational,* and *quasi-rational.* In order to recognize these differences, not only was the invention of game theory necessary, but the operation of the basic mechanisms postulated by game theory had to be proven in different branches of science; and furthermore, a thorough change of attitude consequent to the appearance of Gödel's theorem was required. We can now easily imagine noncausal processes, and we can accept the limits of pure rationality. We must learn to live with these limits, since it was with the aid of logic itself that we have been able to prove their existence.

Jung's idea was a breakthrough in the direction of what became game theory, although this is surely not what he had in mind. Jung's footnote demonstrates how radically new this idea was in contemporary psychology and the world of ideas: "Modern physics has put an end to this strict causality. Now there is only 'statistical probability.' As far back as 1916, I had pointed out the limitations of the causal view in psychology, for which I was heavily censured at that time."

Jung continues:

> The plenitude of life is governed by law and yet not governed by law, rational and yet irrational. Hence reason and the will that is grounded in reason are valid only up to a point. The further we go in the direction selected by reason, the surer we may be that we are excluding the irrational possibilities of life which have just as much right to be lived. It was indeed highly expedient for man to become somewhat more capable of directing his life. It may justly be maintained that the acquisition of reason is the greatest achievement of humanity; but that is not to say that things must or will always continue in that direction.

Freud, Jung, and the other representatives of depth psychology have shown that man's thinking, behavior, and motivations are often fundamentally determined by nonrational elements. Freud, for instance, differentiated between primary processes, which are characterized by the neglect of space, time, and traditional logic, and secondary processes, which are based on the principle of reality and on pure rationality. Both types of processes appear in healthy adults. According to Freud, dreams are the purest example of a primary process. But the dominance of the primary processes is also a common characteristic of different meditative states. The terms *primary* and *secondary*, however, may be misleading. These adjectives reflect the attitude that our conscious thinking is fundamentally rational, but secondary, while the primary processes governed by unconscious, instinctive, and other irrational forces take place under the conscious surface. But are we certain that conscious thinking is truly rational, and that instincts and other unconscious forces are completely irrational?

Logically Isomorphic Tasks

There is a host of experimental data reproduced in my book *Ways of Thinking* that demonstrates that even those problems that can in theory be solved perfectly by logic are not solved in human thought by the methods of pure logic. The chief focus of these experiments is *logically isomorphic tasks*. Two tasks are called logically isomorphic if their formal logical structures are identical. In other words, if we can solve problem *A* by some logical deduction, then we can also solve problem *B* by the very same logical deduction; we have only at the worst to substitute one word for another. Such a substitution can be done completely automatically, for example by using the *replace* function of a word processor. After replacement, we automatically receive the solution of the second task from the solution of the first task. This can be done because formal logic is based on *formal* replacement rules.

Let us have a look at a simple, logically isomorphic pair of tasks. Here is the first task: Let us assume that (1) pelicans eat fish, and (2) Paula eats only lizards. Can we conclude that Paula is not a pelican?

Here is the second task: Let us assume that (1) Healthy people eat mashed yeast, and (2) Paula eats only bean sprouts. Can we conclude that Paula is not healthy?

The two problems are logically isomorphic, because if we solve one, we have automatically solved the other. In the first case, our reasoning is as follows: Suppose Paula were a pelican. Then by (1) she would eat fish. But by (2) Paula eats only lizards. Therefore Paula does not eat fish, and so she cannot be a pelican. If this deduction is logically correct, then we may substitute *pelicans* everywhere with *healthy people, fish* with *mashed yeast,* and *lizards* with *bean sprouts* to conclude that Paula is not healthy.

The two tasks are logically equally easy or difficult; we cannot say that one is more difficult than the other because its solution is more complex or that it requires deeper thought. Yet people usually solve the first problem more quickly and correctly, while they solve the second problem more slowly, and often incorrectly. Thus, the two tasks are not equally difficult psychologically. The reason for this is that the two statements are different for our everyday thinking even if their logical structures are exactly the same. Our attitude in accepting the starting hypothesis "pelicans eat fish" differs from that of accepting that "healthy people eat mashed yeast." The first statement is general knowledge, while the second is at best a working hypothesis. In the latter case we can easily be misled if we rely on our everyday experience rather than apply the strict laws of logic.

As I wrote in *Ways of Thinking:*

As time passed, a kind of mass sport developed among psychologists working in this field: Let us construct logically isomorphic problems whose solution times differ from each other as much as possible. The aim is to compose tasks that are solved by many people sooner or later, although some might solve them much more slowly than others. This game was not engaged in for its own sake, although it is exciting just to engage in sports and break records. But the best results are achieved by those who have the best grasp of how man's reasoning mechanism can be triggered or hindered. The investigators succeeded in figuring out pairs of (I repeat, logically totally isomorphic) problems where the solution time of the more difficult task was 10 to 12 times that of the easier one.

Not only logical tasks demonstrate this phenomenon. At the end of Chapter 3 we presented two versions of the prisoner's dilemma that were logically isomorphic to the original prisoner's dilemma. We have seen that the three logically identical situations elicited different degrees of cooperation from the experimental subjects. The three logically isomorphic games proved to be significantly different psychologically.

The logically isomorphic rephrasing of the situation is an operation that is carried out completely within the framework of rationality. However, the fact that such changes could cause significant changes in the behavior of the subjects demonstrates that on the one hand, it is not only rational elements that take part in our thinking, and on the other hand, our thinking can be influenced by the means of rationality.

On the Role of Rationality

It is often the case in mathematics that the logically isomorphic rephrasing of a problem opens up the way to its solution. Mathematicians consider it very elegant if a deeply hidden relationship can be found between two seemingly very distant mathematical areas. Discovering such relationships has often led to revolutionary discoveries. Methods developed in one area may give an impetus to the other area, which seemed to be in a rut, and sometimes well-tried methods of the one area prove their real strength in another field. Such was the case when mathematicians began to apply the methods of complex analysis to number theory. It turned out that the most suitable method for studying the distribution of prime numbers among the integers was provided by functions of a complex variable.

Mathematics is by its nature a purely rational science. A mathematical proof is a series of logical steps in which there is no place whatsoever for subjective considerations. Nevertheless, when a mathematician is thinking about a problem or is searching for new mathematical truths, he or she is not thinking deductively. The great French mathematician Jacques Hadamard, for instance, wrote a long essay on the nature of mathematical invention, in which he described strange, obscure images that in his mind accompanied mathematical concepts. These images were completely unintelligible and confusing even for another mathematician. If Hadamard had not been able to transmute the relationships revealed to him in these images into the form of mathematical proofs, nobody would be interested in the images themselves. Hadamard reports that Albert Einstein also conceptualized with the aid of such "more or less clear images, which could be combined and reproduced at will."

Henri Poincaré, one of the founders of the mathematical basis of the theory of relativity, after having stated that "mathematics is a language by which no indistinct, obscure, and indefinite things can be expressed," describes how differently he perceived mathematical problems depending on whether he was thinking in French or English. If only pure logic were in operation here, language would make no difference. And it really does make no difference at the moment when thought is realized in its usual mathematical form: definition, theorem, proof.

As we saw at the end of Chapter 4, there are many concepts of rationality, all of them justified, and there is no rational means to single out any one of them and consider it *"the"* concept of rationality. Yet at the same time, some characteristics can be found that are common to all of them. Perhaps it is not the formal logical elements that are primarily to be emphasized, but that *in the case of knowledge reached by a rational method, we not only know what we know, but we also know exactly how we know it.* In these cases what we know can always be deduced from a few simple, basic assumptions and a few simple rules of deduction, although the deduction itself may be very long and complex.

The results of experiments with logically isomorphic tasks show that man does not make use of rational thinking very easily, and even less automatically. However, the same is true of meditative thinking. As we saw in Chapter 12, complicated techniques may lead to deep meditation. But it is more difficult to state how we know what we have acquired through meditation, and yet it may be valuable knowl-

edge about real phenomena of the real world. It is clearly not sufficient to say, "I know what I know because as I plucked the daisy's last
petal I was saying, 'He loves me,'" or, "The coin showed *heads*," or,
"My left index finger rose by itself." This would furthermore not be
enough because the petal, the coin, and the ideomotor technique
merely provided a means for meditative thinking to take place and
made the expression of the outcome of meditation possible.

The main achievement of rational thinking is that it not only leads
to knowledge itself, but it can also demonstrate clearly how that
knowledge was attained. This is why knowledge that can be expressed rationally can be passed on to others clearly and definitively.
People who know and can apply the rules of some type of rational
system (such as logic) can be taught a great portion of the world's ac-
cumulated knowledge quite easily—particularly those parts that can
be expressed in a rational form. The results of science are such
knowledge, and that is no mean accomplishment. Perhaps this is the
reason why the rational modes of thinking have become so generally
accepted in our era.

Although rationality is a very effective tool of communication, it is
less effective as a tool of thinking and even less so as a general tool of
getting to know the world. For the sake of communication we usually strive to express our thoughts and ideas in a rational form, even
if we do not arrive at them by purely rational methods. Nevertheless,
some people think more rationally than others, even though they
may not be more intelligent; that is, they may not be more successful
in recognizing the truths of the world. On the other hand, those who
use other means for thinking also behave very rationally when they
jump out of the way of a swiftly moving bus. Thus, there must exist
some kind of rationality in all of us. The question remains whether
this is the result of rational *thinking* or of something else.

Descartes's Error

Neurobiologists consider it their primary task to localize groups of
brain cells responsible for specific functions. When this problem has
been more or less solved, the next task will be to understand the har-
monious functioning and interactions of these cells.

This strictly reductionist, purely scientific approach has been very successful, from the discovery of the speech centers to the localization of the brain areas responsible for different emotions. The results have not been unambiguous. For example, some people who through disease or trauma have lost some, or even all, of their speech centers have nevertheless learned to speak to some extent, which means that other brain areas were able more or less to take over the functions of the speech center. This does not diminish the significance of our knowledge of which brain areas are responsible for the function of speech in the healthy brain. The problems raised by extreme cases can be put on ice until we know the functioning of the brain more exactly and until we learn which brain areas can take over the functions of which other areas and how such exchanges are accomplished.

However, the areas for conscious and rational thinking have not as yet been satisfactorily localized. In certain cases of brain damage, temporary or permanent disorders of consciousness appear, but these structures have not yet been precisely mapped out. It is even more perplexing that these injuries often accompany disorders in emotional responses. Neurobiologists have begun to suspect that rational thinking and the ability to make decisions are inseparably connected with emotions, or at least with the so-called secondary emotions. Those emotional manifestations are called *secondary* that do not take place upon a real event but upon an imagined one. As the Portuguese–American neurobiologist Antonio Damasio has said, "Nature, with its tinkerish knack for economy, did not select independent mechanisms for expressing primary and secondary emotions. It simply allowed secondary emotions to be expressed by the same channel already prepared to convey primary emotions."

In his book *Descartes' Error*, Damasio describes clever experiments in which he utilizes just this phenomenon in studying the relationship between emotions and rational thinking. According to Damasio's hypothesis, certain learned relationships between the choices we make and the consequent good or bad results elicit secondary emotions. Damasio calls these particular secondary emotions *somatic markers*. According to his hypothesis, these somatic markers really exist, and if they really exist, then they govern our decisions by warning us through certain visceral good or bad feelings, such as that the possibility we are considering seems promising, or perhaps very dangerous.

Since in any case we do not possess the capacity to analyze the consequences of the consequences of all available possibilities, the somatic markers may serve a very useful purpose by narrowing down the area of possibilities to be investigated, not only in the area of direct possibilities, but in the later stages of thinking, too, when we consider the consequences of the consequences. The somatic markers always signal at the moment when the situation arising from our thoughts requires them to do so, namely, when the learned relationships that they symbolize are apparent. In this way they help in narrowing the field of alternatives to a manageable size. They influence our thought processes by eliciting visceral sensations, inspiring us to avoid courses of action that are "bad" for us and to follow courses that are "good" without our understanding why some courses might be considered good, others bad.

Damasio conducted several experiments to prove his hypotheses. In his perhaps most clever experiment the experimental subjects (brain-damaged patients and healthy control subjects) engaged in a little wager. They had to draw a card from one of four different decks. Before each draw the subjects had to decide from which of the decks they were going to draw. If they chose the first or second deck, they would receive $50, but if they drew certain of the cards from these decks they would suffer a loss in the range of fifty to two hundred dollars. Choosing the third or fourth deck netted $100, but quite a few of the cards there resulted in a severe loss to the player, sometimes as much as $1,000. The subjects' emotional reactions were measured by galvanic skin response.

The majority of the subjects quickly learned that although the last two decks brought a greater initial profit, they were also more dangerous. They soon exhibited secondary emotions just by thinking of drawing from the more dangerous decks. These emotions always appeared when they selected one of the latter two decks and also even if the possibility only *occurred* to them. In addition, they also exhibited primary emotions upon winning or losing.

Some patients with particular types of brain damage could not learn the rational strategy, namely, avoiding the last two decks. In these subjects, no secondary emotions appeared when they imagined the selection of the two dangerous decks, although other (primary and secondary) emotions did appear in them. Their thinking functioned well when they had to solve other kinds of tasks (like arithmetic problems). Thus their problem did not lie simply in an inability to think rationally. In fact, at the end of the game they could tell perfectly which two decks were the "bad ones." Their emotional and rational abilities functioned well in themselves. Only these special secondary emotions, the somatic markers, were missing, and in the absence of these markers they could not learn rational behavior.

Damasio gave the striking title *Descartes' Error* to his book because his experiments proved that thinking and bodily functioning are very closely related. Without the visceral sensations elicited by the somatic markers, rational behavior becomes impossible. This contradicts to some degree Descartes's idea that mind and body are a duality of distinctly functioning entities.

The *functioning* of the somatic markers can be considered as completely rational. They condense existing experiential relationships into a simple, visceral response. If the task of a fully rational thinker (say, an engineer) were to develop such markers, he would solve the problem exactly this way. Our thinking, however, does not *use* these markers in a purely rational way, for the relationship between the learned relations represented by the somatic markers and the task to be solved is not understood logically. Our thinking often revises them. Nevertheless, the healthy experimental subjects sometimes selected one of the dangerous decks, despite the already developed somatic markers. Sometimes, all of us leave the beaten track—and we are right to do so!

Somatic markers are not employed purely rationally in thinking, but neither are they used irrationally. They mark relationships whose real existence is highly probable, and thus it is reasonable to take them into consideration to some extent. It is their further use that they limit the alternatives to be investigated—which we badly need in view of our limited reasoning capacity. The *use* of somatic markers belongs among the quasi-rational tools of our thinking.

Where Does Rationality Reside in Us?

Probably, somatic markers constitute only one of the bodily functions that fundamentally influence our thinking. It may well be that our instincts, desires, bodily needs, and even our laws and traditions have a similar effect on our thinking (thinking that we think should be rational), although as of yet we have no clear experimental evidence of this (as opposed to that of somatic markers). Anyone who ever has spent hours or days thinking about a difficult mathematical problem has probably felt similar internal, visceral sensations, just as our shepherdess experienced in trying to decide by counting petals whether *he loves me, he loves me not.*

Representatives of depth psychology like Freud and Jung considered conscious thinking as a purely rational, although secondary, process, while they considered unconscious, "primary," processes irrational. According to their views, the dominance of the latter causes irrational behavior. The discovery of the somatic markers challenges this attitude, which was based on the tradition of the rationalist philosophers (like Descartes). The discovery of the unconscious part of the psyche remains an enduring accomplishment of Freud and Jung, as does the recognition that there are many forces within us that are not known by our conscious thought, and thus we cannot consciously take them into consideration.

Game theory has also opened the way toward another kind of interpretation. Again, events take place at two levels. The first is the level of pure strategies, which are—in themselves—reasonable, regular, and possible methods of playing, although most of the time they cannot be effective enough by themselves. The second level is that of mixed strategies, which determine the proportions in choosing the various pure strategies.

The elements of the *first level* are the consistent ways of playing, those that can be predicted exactly and do not depend upon chance at all. They would be considered as very rational in themselves if the totality of the game were not taken into consideration. Like somatic markers, they are *unconditional strategies.*

The *second level* can be considered perfectly rational from the aspect of the whole game, provided that we can really realize the optimal mixed strategy. This level, however, is completely *based on chance*, which is an irrational method according to Jung. As we have seen, in reality we do not realize this level by strict probabilistic methods such as throwing dice, although we do not employ fully rational means, either. Is it possible that this level is the level of conscious thinking? Is this the level that Freud called secondary thinking, and do pure strategies correspond to primary processes?

According to the emerging picture, *unconscious processes are those that are fully rational, while it is conscious thinking that is not completely rational.* The latter is quasi-rational at best.

This is supported by the evidence from experiments on logically isomorphic tasks. The results are generally explained in terms of everyday life: It is easier to solve those tasks that are more closely related to our everyday experience, to which we can connect our "mental models" more easily. Experimental evidence really demonstrates this. With our knowledge of the somatic markers, however, we can add that our unconscious somatic markers may also promote solving problems closely related to our everyday experience or at least may direct attention to promising ideas or those likely to lead to dead ends.

Our somatic markers certainly sometimes mislead us in the process of logical deduction, as in the example of health through mashed yeast. We may assume that ingesting mashed yeast is no more healthful than dining on bean sprouts, while in the example we assumed that only the former brought good health. In the end, by the aid of logic, we think about what a world would be like in which bean sprouts are actually not good for you. After all, the world could be like this or like that. However, every conclusion drawn from this (for example, eating yogurt is bad for your health) is valid only in the abstract world of logic. Everyday life does not function this way. Our somatic markers point this out strictly and consistently. Perhaps it is not really disadvantageous that pure rationalism is not exclusive in our thinking.

The British physicist Roger Penrose arrived at very similar conclusions from a totally different line of reasoning. In his book *The Emperor's New Mind* he asked how realistic is the belief of the followers of so-called strong artificial intelligence, namely, that human thinking can be described by purely rational symbolic manipulation (and thus executable by computers). He concludes this at the end of a very complex argument in quantum physics and the theory of algorithms: "It is ironic that the views that I am putting forward here represent almost a reversal of some others that I have frequently heard. Often it is argued that it is the *conscious* mind that behaves in the 'rational' way that we can understand, whereas it is the unconscious that is mysterious."

When we introduced game theory through the analogy of the schizophrenic snail, we remarked that perhaps our analogy was too psychological. Opposing forces operate within the snail that would like to realize their own aims. These opposing forces may correspond with its conflicting instincts, desires, and needs. In our present terminology, we could say that they develop their own somatic markers for their own purposes. If the snail took into consideration all of its somatic markers all of the time, it would never arrive at a decision, for these individually totally rational markers would always contradict one another. This would drive the snail crazy. Therefore, the snail must decide at every moment which markers to neglect. If, however, it neglected some of the markers consistently, it might not fulfill an essential function.

It is a great achievement of the scholars of depth psychology that they discovered the unconscious forces operating within us. These discoveries are still valid, and their significance will not be lessened if by any chance it turns out that it is not the "dark" forces of the unconscious that are irrational in us but that it is their very consistent, severe rationality that is alien, unintelligible, and irrational to our not at all purely rational conscious thinking.

Games People Play

There is a special kind of everyday, nonrational game that was discovered and described by Eric Berne in his book *Games People Play*. The analyses given by Berne are radically different from the approach of game theory. Nevertheless, we can find quite a few similarities.

"Schlemiel" is a typical game people play that may take the following course. Mr. White ("schlemiel") spills a glass of whiskey on the evening dress of his hostess. Mr. Black, the host, is boiling with rage, but the situation prevents him from showing it. He might even feel that if he showed it, White would "win" by achieving a pleasure, namely, that of taking offense at the host, of considering him petty and heartless. After Black suppresses his anger, White apologizes. Black generously forgives him, but White also thereby scores a point.

He has learned that he, a schlemiel to the core, may behave like one with impunity.

Berne continues the description of the situation (p. 114):

> White then proceeds to inflict other damage on Black's property. He breaks things, spills things, and makes messes of various kinds. After the cigarette burn in the tablecloth, the chair leg through the lace curtain and the gravy on the rug, White's Child is exhilarated because he has enjoyed himself in carrying out these procedures, for all of which he has been forgiven, while Black has made a gratifying display of suffering self-control. Thus both of them profit from an unfortunate situation, and Black is not necessarily anxious to terminate the friendship.

Berne describes many other games people play in marriages, office relationships, pubs, or even on the psychiatrist's couch. It is very easy to get entangled in such situations, since to some extent they are good for both participants, both parties finding some psychological satisfaction in them. Nevertheless, these games operate very similarly to the traps we examined earlier, like the dollar auction. After a while, both players are fed up with the game, yet it is very difficult to exit. In order to quit, one would have to give up the small, but certain and reliable, psychological gain and would have to leave a system of behavior to which one had become accustomed.

Quitting, like the trap situations of game theory, can be achieved only through finding a higher-order principle. Freud, for instance, saw clearly that psychoanalysis itself may easily become a kind of game people play, and one may thus fall into one's own trap. Therefore, he often emphasized that he, Freud, was not himself a Freudian.

Berne remarks that while mathematical game theory assumes completely rational players, he deals with nonrational and even irrational games, which he thinks are more realistic. This can be debated, perhaps. Both worlds of ideas study model situations. Berne analyzes his models with the language and methods of psychiatry and studies psychological gains and possibilities of rising above the games. John von Neumann analyzed his models with the language and methods of mathematics. He talked about the objectivity of what is won or lost, and thus he could say something about the concept of rationality as well. For von Neumann, leaving the system of "yes or no" rationality meant the development of the concept of mixed strategies, and we have seen the significance of this at many places. Berne studies the antithesis of the games, which he describes as quitting a game, and in general, he finds these antitheses in three deeply human abilities: consciousness, spontaneity, and intimacy. Of these, spontaneity stands the closest to game theory, but all three guiding principles of Berne's are excellent and general quasi-rational strategies of problem-solving.

Berne describes the antithesis of "Schlemiel" as follows:

> After White says "I am sorry," Black, instead of muttering "It's okay," says, 'Tonight you can embarrass my wife, ruin the furniture, and wreck the rug, but please don't say 'I am sorry.'" Here, Black switches from being a forgiving Parent to being an objective Adult who takes the full responsibility for having invited White in the first place.

This way Black ends the game and offers White the opportunity of quitting as well. It is a dangerous and courageous step to exit the vicious circle of "Child–Adult–Parent" so familiar to all of us, but this is the only chance—particularly in deeply rooted, difficult games.

The American psychiatrist Paul Watzlawick calls this kind of quitting from an entrenched situation that may perhaps lead to tragedy *second-order change*. He gave the following example:

> During one of the many nineteenth-century riots in Paris the commander of an army detachment received orders to clear a city square by firing at the *canaille* (rabble). He commanded his soldiers to take up firing positions, their rifles leveled at the crowd, and as a ghastly silence descended he drew his sword and shouted at the top of his lungs: "Mesdames, m'sieurs, I have orders to fire at the *canaille*. But as I see a great number of honest, respectable citizens before me, I request that they leave so that I can safely shoot the *canaille*." The square was empty in a few minutes. (*Change,* p. 81)

This is how Watzlawick summarizes the psychological essence of the situation and its clever resolution:

> The officer is faced with a threatening crowd. In typical first-order change fashion he has instructions to oppose hostility with counterhostility, with more of the same. Since his men are armed and the crowd is not, there is little doubt that "more of the same" will succeed. But in the wider context this change would not only be no change, it would further inflame the existing turmoil. Through his intervention the officer effects a second-order change—he takes the situation outside the frame that up to that moment contained both him and the crowd; he *reframes* it in a way acceptable to everyone involved, and with

this reframing both the original threat and its threatened "solution" can safely be abandoned. (*Change*, p. 82)

The solution of the officer is typically quasi-rational. It cannot be rational, because in no way does it follow from the hostile logic of the given situation. He could not have deduced such a solution from his professional studies. Yet the intelligent nature of the solution cannot be doubted. It is also possible that the illuminating thought occurred to him in the absorbed, meditative moment of drawing his sword. Most certainly, he did not behave like this in every situation, for otherwise he would not have become an army officer. However, this possibility was also present in the mixed strategy of his consciousness, and his quasi-rational consciousness formed this very solution in the given moment.

Further Aspects of Games

Games people play were not discovered by game theory. Berne considered it necessary to mention in the preface of his book, published in 1962, only that the transactional analysis of games must be clearly differentiated from its quickly developing brother, mathematical game theory. He is justified in settling this matter this way, because the methods and problems are fundamentally different. Just as there is no single, well-definable road of rationality, mathematical game theory is not the only successful method of investigating games. There also exist many fertile approaches beyond Berne's analysis of games.

For example, the Hungarian neurobiologist Endre Grastyán and his colleagues have demonstrated in animal experiments that a number of activities that are totally unnecessary for self-preservation cause physiological changes that are essential for the health of the animals. Here is an aspect of play that is totally neglected by game theory, although it is fundamental. Grastyán's results render it highly probable that it is the *absolute pointless nature of play* (pointless for survival, that is) that causes the physiological changes essential for health. As Grastyán states, "One cannot be forced to play by punishment, nor can playing be rewarded outside its own sphere. Playing is sustained by its own internal rewards. . . . A monkey playing with a puzzle stops playing as soon as it is rewarded for success. . . . Real play takes place in the sphere of pure ethics; it is the most ethical activity, because the source of joy is equivalent to the realization of the rule." Our game theory had nothing to say about these very important questions.

Philosophers study the philosophical and cultural–historical significance of play from another perspective and by different methods. For instance, they will not be, as we were, satisfied with the intuitive treatment of games, but even an exact mathematical definition could not sufficiently describe for them the fundamental problems inherent in the concept of "game." Johan Huizinga (already cited in Chapter 13, when we talked about game spoilers) devoted an eighteen-page chapter to expressions corresponding to *game* in different languages, from Greek to Chinese, from German to Sanskrit. Let me cite a paragraph from another chapter in Huizinga's book to illustrate his tone:

> With games one recognizes, willy-nilly, the mind, the spirit. For games are not substance from which one's being may arise. Already in the animal world it breaks through the bounds of physical existence. From the point of view of a deterministic world of pure forces it is in the fullest sense a superabundance, something extraneous. Only through the inflowing of mind, which abolishes absolute determinacy, is the presence of play made possible, thinkable, and understandable. The presence of play verifies repeatedly, and indeed in the highest sense, the superlogical nature of our place in the cosmos. Animals can play, and thus even they are more than mechanical objects. We play, and we know that we play, and thus we are more than reasoning beings, for play is unreasonable.

Game theory would have nothing to be ashamed of if it had similarly nothing to say about the questions that Huizinga dwells on, as it has nothing to say about the physiological significance of playing. It is all the more interesting that in the last chapter of the book we are going to arrive at conclusions that are in many respects in accordance with Huizinga's ideas. This also shows the power of scientific thinking, and, on the other hand, that we can arrive at similar conclusions by taking very different paths.

15

There Are Many Ways to Nirvana

In the most different ways we are all alike.

In an old fable, an emperor ordered his wise men to summarize the essence of wisdom in a single book. After years of debate, the wise men produced a weighty tome. But during this time the emperor had grown old, and he realized to his dismay that he would have no time to digest the leviathan his wise men had produced. He decreed that they should summarize the essence of wisdom as concisely as possible. By the time the required compendium was complete, the emperor had grown so old and frail that he knew that he would have no time to understand even this brief book. He therefore commanded his wisest wise man to summarize the content of this book in a single sentence. The wise man thought and thought, and after thinking for a long time, he sent word to the emperor's chamberlain that he had found the sentence the emperor was seeking. He was led into the royal bedchamber, where the emperor lay dying. "Speak, O wisest of all my wise men," said the emperor. "What is the essence of wisdom that you have labored so long to distill?" The wise man replied, "Your Highness, the essence of wisdom, the essence of all knowledge, is this: The world is complicated."

Were I ever to find myself in a similar situation, compelled to summarize the most important lesson from my studies in psychology, I would say, *We are diverse.* If by the emperor's grace I were allowed to add another sentence, I would also mention the conclusions drawn from general and experimental psychology: *It is amazing how similar we are.* If I were to summarize in a single sentence the subject of the present book, I would probably say, *Its subject is the compatibility of these two conclusions.*

That is what I would probably say, but I am not sure. Perhaps, standing in awe before the royal presence, my momentary mood would dictate that I emphasize something else, like the role of game theory in changing the attitudes of various branches of science, the diversity of rationality, or the quasi-rational nature of meditative cognition. I use mixed strategies, after all, just as we all do. This is the basis of both our diversity and our sameness. We all think by means of mixed strategies; in this respect we are basically alike. Yet the pure strategies underlying the mixed strategies and the ways of mixing them are totally individual; hence the basic diversity of our habits of mind and our personalities.

There are many ways of getting to know the world. Science has shown how effective pure rational thinking can be. In reaching internal harmony, balance, and peace, however, meditative methods like those of Eastern thought provide more effective tools. Game theory has helped us to understand how several opposing strategies (selfish gene and group selection, rationality, and intuition) can exist simultaneously in nature's great game, and how they can lead to a higher level of equilibrium both in the world and in human thinking. At the same time, game theory has given us a better understanding of the nature of *individual* game strategies.

The Nature of Rational Cognition

Albert Einstein once said, "It is the least understandable feature of the world that it is understandable." This was Einstein's scientific credo. In his book *Mein Weltbild* (*My Worldview*) he discusses in detail how his unshakable belief that the world can be understood led him to all his discoveries.

For Einstein, *why* the world is understandable falls outside the sphere of what can be understood. This is not a question for science. That the world *is* understandable, or at least that some phenomena can be apprehended, is self-evident to every scientist. That is the *definition* of "scientist." If someone does not feel clearly this intelligibility, then other scientists will not consider this individual a colleague. And if they should enter into debate over this question, the scientists will not consider it a scientific debate, but a discussion with an outsider, an ignoramus—perhaps even a struggle against the "forces of darkness." For scientists, the starting point in getting to know the world is unambiguous. Any other point of view is opposed to human rationality.

Scientists set out exclusively from experimental observations and precisely formulated hypotheses (models), and they draw conclusions by *pure reason*, preferably that of formal logic. It is the essence of the scientific method that *it does not contain any subjective element*. It follows from the rules of this game that if two individuals start from the same experimental evidence and original hypotheses, they must *theoretically* arrive at the same conclusions, provided that their intellectual capacities make this possible.

Nonetheless, there are heated debates in the scientific community. The debates, however, are generally not about the acceptability of particular scientific deductions, but about the correctness of the hypotheses. If the experimental facts themselves are questionable or the process of reasoning is doubtful, the debates are usually quickly settled. Every scientist strongly believes that such questions have objective solutions, so consensus is soon reached. It is around the starting hypotheses, theories, and models that long and difficult scientific debates arise. But even in these cases scientific debates develop only if the methods of all of the competing theories conform to the scientific method, if the theory can be phrased in purely rational terms so that it can be treated—at least in principle—within the framework of a formal system.

At least in principle. In practice, however, it is very difficult, if not impossible, to realize all this. Man makes use of the excellent methods provided by formal logic only with great difficulty. Scientists, too, think intuitively even about questions whose solutions are eventually couched in terms that are purely rational. Furthermore, the determination that a particular chain of reasoning is really scientific is often based on intuition. Schrödinger's equation was quickly accepted by physicists, although its derivation cannot be accepted on purely logical grounds. Perhaps this is the *very reason* why it could offer something basically new, a set of axioms that was radically different from its predecessors. However, the new theory fit well into the intuition of those physicists to whom this idea had not occurred. It fit into existing physical intuition, but it also changed that intuition radically. This is the essence of every brilliant idea. It is the least understandable feature of brilliant ideas, too, that they are understandable.

It is the essence of *scientific intuition* that when a scientist is thinking intuitively, he or she feels every moment that if the conclusion is correct, then it will be possible to express it in purely rational terms. This feeling governs scientific intuition and differentiates it from other forms of intuition. By the time one becomes a scientist, this intuition has become firmly developed. When scientists have discussions with one another, they permit themselves to express their thoughts in ways other than the slow and difficult reasoning of pure rationality, ways that would have no place in serious scientific publications. All of the parties in the discussion sense exactly whether an interlocutor is thinking scientifically or not. This sense is not based in some formal system but is rather a kind of "musical ear." Scientists sense a line of thought that has no chance ever of being expressed in purely rational terms as somehow dissonant or out of tune. This intuition is similar to the one we Central Europeans use when turning the dial on a radio. Suddenly, we prick up our ears, because although we did not understand a word, we are sure we heard a station where our mother tongue was spoken.

The scientific method can help us with only a very limited range of questions, and this limitation belongs to the essence of the method.

If the universe of discourse were not limited, the question would be unlikely to yield an answer that could be expressed along purely rational lines. Paradoxically, in many cases these very limited questions have led to highly general answers, like the principle of conservation of energy. Furthermore, very precisely defined questions have led to results like Gödel's theorem, which showed the *general* limits of formal systems. We saw at the end of Chapter 4 that based on Gödel's theorem, game theory, working with purely rational methods, could illuminate the limits of even the concept of rationality, that no single concept of rationality can express all the possible rationalities in the world. It is possible to create a game for every concept of rationality in which the given concept would evidently lead to an irrational result, although another type of rationality might prove successful.

This does not call into question the usefulness of rationality as a method; in fact, it even reinforces it. Such general results, results that show a method's own limitations, could have been achieved only by the methods of rationality. This is reassuring: Our belief in the understandability of the world can discover its own limitations by its own methods, and thus we can be reconciled to the fact that we are condemned to switching eternally from one system of thought to another.

The Nature of Mystical Cognition

Meister Eckhart, the outstanding representative of German mysticism in the Middle Ages, wrote the following at the end of a long and profound argument:

> When I flowed forth from God, creatures said: "He is a god!" This, however, did not make me blessed, for it indicates that I, too, am a creature. In bursting forth, however, when I shall be free within God's will and free, therefore of the will of god, and all his works, and even of god himself, then I shall rise above all creature kind, and I shall be neither god nor creature, but I shall be what I was once, now, and forevermore. I shall thus receive an impulse which shall raise me above the angels. With this impulse, I receive wealth so great that I could never again be satisfied with a god, or anything that is a god's, nor with any divine activities, for in bursting forth I discover that God and I are One. Now I am what I was and I neither add to nor subtract from anything, for I am the unmoved Mover, that moves all things. Here, then, a god may find no "place" in man, for by his poverty the man achieves the being that was always

his and shall remain his eternally. Here, too, God is identical with the spirit and
that is the most intimate poverty discoverable. (pp. 231–232)

Then Meister Eckhart continued: "If anyone does not understand
this discourse, let him not worry about that, for if he does not find
this truth in himself he cannot understand what I have said—for it is
a discovered truth which comes immediately from the heart of God."

Rational cognition is based on the belief that the world *can be understood in full*. Mystic cognition is based on the belief that the world
can be experienced in full. The rational view of the world is based exclusively on empirical facts that can be experienced by our physical
sense organs, and all other knowledge is left to correct reasoning.
According to the mystic view of the world, if one becomes sufficiently clear spiritually, one can then acquire the ability to experience union with the world and no longer distinguish self from nonself, the world within from the world without. There is no place for
logic and rationality here, because logic and rationality are for discovering and organizing differences in the world, and any differentiation removes one from the experience of the profound unity of the
world. Those who can reach this mystic state can know the world
through identification, or, to be more exact, through becoming one
with the world. The belief in the basic unity of the world and in the
possibility of experiencing this unity lies at the foundation of all
mysticism.

In Saint-Exupéry's tale *The Little Prince*, the fox tamed by the prince
reveals a great secret: "One can see well only with the heart. Truly important things are invisible to the eye." The rational view of the world
fully agrees with the fox's second sentence, but it considers this *the
very explanation* for why the power of *reason* is needed, why correct
conclusions are arrived at as the result of complex logical reasoning.
According to the mystic view of the world, *neither the heart nor the
mind* can help in truly getting to know the world. In fact, they are actually obstacles in this. In order to experience the mystic unity of the
world, we have to develop an extra "sense organ." This sense organ
can be only the totality of what we are, for if it were a real "organ," a
part of us, it would lead again to differentiation. Experiencing the
mystic unity of the world is an extremely special state of consciousness. Although the existence of such a state of consciousness does not

follow logically from anything, neither can it be deduced by any rational means that such a state cannot exist.

It is not a logical consequence of the mystic view of the world that mystical experience has to be linked with religious experience. Most of the time it is, but this is not a necessity. The experiences reported by the Eastern mystics are amazingly similar to those of Meister Eckhart, although the religious backgrounds of these mystics are radically different not only from that of Meister Eckhart, but often from one another. Sometimes such a religious link is deliberately excluded, as in the case of the Zen master Dógen Zenji, who happens to have been a contemporary of Meister Eckhart.

In Zen Buddhism, the peak experience of the complete unity with the world is called *satori,* or *enlightenment.* This experience cannot be described by words, for it is the essence of enlightenment that one permanently and completely goes beyond words, for words are the means of differentiation. Enlightenment is a highly ecstatic experience, one that radically alters one's view of the world.

Other enlightened masters are able to judge whether a person has attained satori, and there is generally no disagreement among them in this. They do not make this judgment on the basis of external cues, just as scientists do not decide on the basis of a formal system whether or not somebody is thinking scientifically. Rather, enlightened Zen masters have a kind of "ear for Zen." They can feel a certain pure harmony in every manifestation of an enlightened person. One does not necessarily have to be enlightened to perceive this. We involuntarily feel the internal authenticity of the enlightened person as he orients himself in the world with an internal calmness that cannot be achieved otherwise. Charlatans claiming to be enlightened can at times deceive ordinary persons, but they never deceive the truly enlightened.

Daisetz Teitaro Suzuki, one of the great Zen masters of this century, believes that Meister Eckhart was enlightened, that he attained satori. Only enlightened mystics can authentically make such statements. They sound ill in another's mouth.

In my mouth, too. For me, the path to knowing the world leads through rationality and science, including getting acquainted with other important modes of cognition. We try to understand the behavior of electrons or sticklebacks without trying to *be* electrons or sticklebacks. As the Hungarian poet Attila József wrote: "Skillful though the cat may be, it cannot simultaneously catch the mouse indoors and the mouse outdoors."

Extreme objectivity is a fundamental principle of scientific cognition. Scientific experiments can, in principle, be reproduced by anyone, at any time, although not everybody may have hundreds of millions of dollars to build a superconducting supercollider. Beyond a great deal of learning, a great deal of money is often needed to reproduce scientific results, but if somebody is devoted to the quest, he can convince himself of the truth or falsity of scientific results, no matter what he believes in otherwise.

The means of mystical cognition is the totality of the observer, a special ability of sensation developed within. Therefore, mystical cognition, according to its basic nature, is completely subjective. Thus, science attempts to banish all mysticism from its thinking. Yet, the reproducibility of experiences is fundamental to Eastern mysticism, too. Despite their diverse philosophical and religious backgrounds, the worldviews of the different mystical masters are so similar that one is hardly justified in condemning this mode of cognition as purely subjective and treating its results as exclusively a matter of blind belief. The development of mystical enlightenment is the peak performance of our quasi-rational, meditative methods of cognition.

Fritjof Capra points out in his well-known book *The Tao of Physics* that both scientists and mystics have developed extremely highly developed and refined methods for observing nature that are not accessible to laymen. For a layman, a page of a journal on experimental physics is as mysterious and incomprehensible as a Tibetan mandala. Capra presents many similarities between the worldview of quantum physics and that of Eastern mysticism. The latter undoubtedly has chronological precedence over the former.

Rationality as a Technique of Distancing

Mankind has elaborated various strategies to develop conscious, quasi-rational thinking. *Distancing*, as shown at the end of Chapter 11, is one of the important general methods. Distancing offers a possible *practical* solution to the theoretically still unsolved problem of how man can bring himself into a pure state by observing himself consciously. Distancing enables the psyche that in its natural state is in a mixed state and the psyche that acts as an observer to function simultaneously.

Both conscious thinking and unconscious processes are parts of our psyche. The latter constantly bombard the former with their somatic markers, contributing in this way to maintaining the natural, continuously mixed state. When making a decision, however, one can take into consideration only some of the somatic markers, not all of them. In these instances one has to bring oneself into a pure state, at least for a moment. The very aim of distancing may be to enable the psyche to consider not only the somatic markers themselves, but their relationships, connections, and causes—approaching in this way a truly rational decision.

Almost anything will do for this purpose, anything that can be put between the somatic marker and its cause: plucking petals, projecting images onto an imaginary screen, anything by which we can set the situation in a framework where the "rules of the game" are different from those of the psyche's own natural way of functioning, anything that helps the separation of the observer and the observed. *Logic is also adequate for this purpose:* It is not an automatic, natural means of human thinking. Thus, the framework we choose might as well be purely rational, even scientific.

Some people have been able to develop their logical ability to such a high degree that they are able to apply the rules of logic even in very complicated cases; in fact, by extremely complex logical lines of reasoning they are able to solve problems others can easily solve by everyday intuition. The ability to apply logic has been developed in all of us to a certain extent, the more so, since education has placed rather heavy emphasis on this ability. Perhaps this is not because it is a particularly important means of decision-making, but because it is the most suitable method of conveying information *unambiguously*. Thus, if logic can lead to the solution of a problem in a few steps (perhaps also using the effective aid of somatic markers as well), we will probably use this method, without any particular distancing technique.

Distancing is basically a meditative technique, but sometimes we use logic for the very same purpose. We can stare in wonder at the not-at-all-self-evident relationships of the world, and we can bring ourselves into a psychic state where these relationships become natural to our thinking.

Beyond Rationality

Rationality does not tell us how a problem can be solved rationally if it can be proven that the problem has no rational solution. The form of this question is very similar to the paradoxes that amused us in our childhood. For example, can God create a stone so big that even He cannot lift it? This paradox can be resolved in many ways. For instance, one can create something without having to lift it. But for a religious person the question itself is of no importance, because whatever the answer may be, it has no effect on his belief.

The question of rationality, however, is of utmost importance to one who would like to believe in the power of rationality; for Gödel's theorem proved that sometimes a question phrased within the boundaries of a completely rational deductive system nevertheless has no rational answer. Not because we cannot find it, but because it does not exist.

Science has answered the challenge posed by Gödel's theorem with the opinion that we are now condemned to eternally modifying the systems of science. Beyond rationality there is another, even more powerful, rationality, beyond which there is another, even more powerful, and so on. Put another way, rationality in one field of science is not exactly the same as that in another field. Not because they contradict one another, but because all of them together are too complex for human comprehension. And the number of fields of science is increasing, with no end in sight.

Nature gives a different answer to this problem: *"Rationality is your human concept, just like such concepts as place and speed. It is a good concept. You can build excellent means of cognition with its aid. I do not deny it, but it is not my way. You are left with quasi-rationality as the basic means of your thinking, in all its diversity. If you are sufficiently resourceful, you might perhaps obtain an impression of my real nature (no pun intended) in some mystical way."*

Game theory has broadened the sphere of problems that can be treated by rational means, allowing us to attack problems that had previously resisted all of our efforts. Zero-sum games are a good example of this. With von Neumann's theorem they can be analyzed by the methods of pure rationality. We no longer need such infinite mental loops as "I think that you think that I think that. . . ." The reevaluation of the role of chance—the discovery that rationality may operate probabilistically—was an even more significant consequence of game theory. It enabled us to understand how quasi-rational ways of thinking can approach a higher-order rationality, such as an optimal mixed strategy or an evolutionarily stable strategy. Perhaps we have thereby come closer to understanding the fundamental operating principles of nature. Perhaps. Time will tell.

It is the consequences of Gödel's theorem, however, that are the most breathtaking. Indeed, the number of Gödelian problems has only increased. It has become evident that there are several concepts of rationality and that there are no rational means by which we might differentiate among them. No matter what concept of rationality we accept, that particular rationality can be only one of our many quasi-rational methods.

Gödel's theorem caused great shock waves in the scientific community. The seemingly solid foundations of the scientific worldview were suddenly open to doubt, and it was decades before scientists could accept the theoretical limits of science, and even longer before they could take pride in being able to prove such limits. On the other hand, the Gödelian phenomenon caused hardly a ripple in mystical thought, since one of its main ideas has always been that of transcending logic. Thus mystical thought can be as effective a tool in creating inner harmony as was science in the creation of the foundations that support the achievements of technology.

Perhaps the most significant benefit of game theory was that it helped us to recognize new criteria of rationality that must be built into our quasi-rational thinking strategies as quickly as possible—if we want to avoid the extinction of our species. The sad results of the dollar auction and the prisoner's dilemma show that human thinking is apt to fall into traps that animals easily avoid.

This may not be an advantage for the animals. As far as we know, animals possess no conscious, secondary thought processes that would enable them to adapt to *different kinds* of rationality. Either their primary processes adapt to rationalities that happen to be favored by natural selection, or they eventually become extinct. We are surrounded by animals that have not become extinct—no wonder they prove to be so rational.

It is possible that the circumstances modeled by the dollar auction or prisoner's dilemma became significant only relatively recently. For instance, the problem of environmental pollution is very new on the scale of evolution. If an animal found itself in such a dilemma, it would soon become extinct, and its ecological niche would be taken over by a species whose rational primary processes followed the logic of the new situation. Given enough time, evolution would certainly find the combination of gene selection and group selection leading to a more adaptable species, one that is willing to cooperate more readily. It was given to man, however—being able to think consciously and to know himself—to make the decision to change the mixed strategy he has been using up to now and thus to enable himself to survive in a changing world.

The Two Components of Thinking

It is a fundamental thesis in Zen Buddhism that there can be no definition of Zen. But this is not itself a definition, since Zen could then be defined after all. Nevertheless, the master has to say something when the disciple asks, "What is Zen?" The reply always depends on where the student is on the road to enlightenment. Typical answers are, "Three pounds of flax"; "Oh, you bag of rice!"; "Buddha is this very consciousness"; "It's not consciousness, it's not Buddha"; "The cypress in the yard." Or perhaps the master gives the disciple an unexpected blow with a stick, or with a quick movement cuts off a finger. Whatever is necessary to give the disciple a push toward satori—the last one if he is lucky.

It is as if we were looking at a collection of exercises for understanding the concept of mixed strategy. In this book we have introduced the concept of mixed strategy by another method, the language of rationality. We have talked about the principles of rationality, evolutionary stability, and the categorical imperative, and we have examined the kinds of rationalities that can be reached by the means of mixed strategies, how a mixed strategy can be optimal. What could we have said if the aim of the game had been the total extinction of rationality as a means to open the way to mystical enlightenment? If we had a rational means to analyze this, we would probably arrive again at some kind of mixed strategy, just as Zen masters do.

If the world operates fundamentally according to mixed strategies, then we must be fully alive to this concept if we want to experience mystic identification with the universe, even if our methods (unlike those of game theory) do not include geometry and the theory of probability. Mystic identification is fully alive to *all the mixed strategies of everything in the universe*; from the electron to the sticklebacks, from investors' portfolios of securities to lovers taking off their petals. But this is only words, of course, which can only distance us from mystical knowledge.

Scientific thinking makes every effort to banish mysticism, just as mystical thinking banishes rationality. Each type of cognition attains its peak performance if it can achieve its goals completely. Yet both kinds of cognition are present in everybody. Human thinking as created by nature is composed of the mixed strategy of these two radically opposed methods of cognition. *Humanity itself* cannot exist without these two methods. They exist in individuals in varying proportions, but all of us have both of them.

Even in mystic masters, rational thought reappears after the trance state has passed, and somatic markers become effective again—although what these individuals experience in their mystic state continues to guide their intuition. The masters of rationality in turn often behave remarkably irrationally in their everyday lives, when they think of problems unrelated to their narrow field of science. People who "think with their heart" also use both logic and intuition.

Mystical thinking may be a component of our consciousness that embodies the principle of group selection by its totality, while rationality may embody the principle of gene selection by its analytic function. This is how nature's fundamental guiding principles may appear in human thinking, and this is why both ways of thinking may be fundamental parts of human consciousness.

Game as Essence

In Chapter 12 we left the question open of why the psyche can be thought of as somehow conforming to quantum-mechanical principles. The problem was that although there are many analogies between the mixed strategies of the psyche and the mixed strategies of the elementary particles, there is a fundamental difference. According to quantum mechanics, the behavior of very small elementary particles is nondeterministic, and the closer we get to the macroscopic world, the more the deterministic principles become manifest. In the psyche, however, our simple, everyday decisions are more or less deterministic, while nondeterministic decisions are more characteristic of questions of great importance, those that affect the whole psyche.

Elementary particles are elementary because they cannot be divided into further components. (As soon as one of them can be divided, it ceases to be elementary—just as atoms are no longer considered elementary.) Consciousness is also an entity that cannot be divided into further meaningful components. The neurons of the brain, their complex interconnections and mechanisms of information transmission, are not parts of consciousness; they are only carriers of it—just as water is not a part, only a carrier, of waves. Elementary particles also produce rather stable phenomena, like the energy or, even more, the charge of electrons. However, the particle itself behaves rather nondeterministically in terms of such global human concepts as location or momentum. The psyche also produces rather stable phenomena, such as our everyday, routine decisions. Its nondeterministic behavior is manifested mainly in comprehensive, deeply human questions.

Our words, our desire to create systematic order, can easily mislead us once again. When we talk about the mixed strategies of electrons, we are inclined to forget about the fact that the electrons do not "play" these strategies. It is not the electrons that have mixed strategies that determine their actual position, it is rather that *this mixed strategy is the electron itself!* This is why we could say that in fact the electron has no such thing as a *location*. Furthermore, this is true not only of its location, but also of its momentum and of many other of its general characteristics. Actually, *an electron is the sum of all of these mixed strategies*, at least if we want to understand it by our human concepts. Similarly, it is not our consciousness or psyche that has mixed strategies; rather, the sum of these mixed strategies *is consciousness itself.*

Johan Huizinga, who was cited at the end of the previous chapter, determined by the methods of pure philosophy that "The presence of play verifies repeatedly, and indeed in the highest sense, the super-logical nature of our place in the cosmos." Within the framework of our thought, almost every word means something different to us than to Huizinga. For Huizinga, *game* means a meaningless activity—having no survival value—in a limited world. For us, however, a *game* is any social interaction in which there are possibilities of moves, and after each move the players can evaluate exactly how much each of them has gained. I do not know what Huizinga means by the concept of the *highest sense*, but I am certain that it is not what we mean, namely, mixed strategies, and even the mixed strategies of mixed strategies, applied by nature when she operates gene selection and group selection, or rational and mystical thinking simultaneously. I would consider *the highest sense* as the *optimality of these* (gene selection and group selection, rational and mystical thinking) in the sense of some kind of higher-order rationality (which is not yet known more exactly). Despite the fact that we moved within a totally different system of thought from Huizinga's, our conclusions are rather similar. The great game of nature and human cognition points way beyond every kind of rationality we have known so far, especially beyond its very special form called logic.

Nirvana

Most Eastern religions are based on a belief in reincarnation. According to these beliefs, the aim and chief source of happiness in a person's life is attaining perfection, a state in which the soul becomes completely free of the constraints of worldly desires. This state is called nirvana, and the final goal of every soul is to reach nirvana. According to the Eastern worldview, human souls are governed by nirvana. It would be misleading to say that they are governed by the *desire to reach* nirvana, since the soul must extinguish all of its desires in order to reach nirvana. It would be more exact to say in our terminology that nirvana governs the soul as *a kind of natural force*, just as according to scientific thinking the motion of the celestial bodies is governed by gravity, while the development of species is governed by evolution.

If a soul cannot attain nirvana in one lifetime, it will be reborn. Since nirvana governs the soul, the chief events of one's life are determined by the ethically unsolved or erroneously solved events of a previous life, and one has been reborn in order that these things be rectified, so that the soul might come closer to nirvana.

Opinions vary as to what the nirvana of Eastern religions really is. The etymology of the word lies in the notion of blowing out a fire, or the act of extinguishing a lamp. In Hinduism, nirvana implies a mystic union with divinity and thus has no connotation of extinguishment or annihilation. In Buddhism, nirvana means perfect annihilation. There are many interpretations of the concept within both religions. But it is a common element in every interpretation that nirvana means *becoming free of the compulsion of reincarnation.* It is the end of earthly suffering. If someone's soul has attained nirvana, that person may continue to live on Earth for a while, but he will not be reborn again. Therefore, psychiatrists in the East do not encounter much clinical depression associated with fear of death, but they do see a similar phenomenon, depression arising from the realization that one is highly unlikely to attain nirvana in this life and so reincarnation is unavoidable (the Hungarian psychologist Péter Popper calls this "fear of life"). And the patient still has to get through this life in such a way as to leave as little as possible in disarray to be redeemed in the next life.

Eastern and Western religions thus differ greatly from each other. An adherent of a Western religion has to arrange for the eternal future of his soul in the course of a single life. Belief is indispensable, for otherwise the soul will be condemned to eternal punishment. In Eastern religions the soul may reach, or at least approach, nirvana even in the absence of belief. Therefore, religious tolerance is widespread in the East. Another person's soul may be governed differently by nirvana; it may be weighed down under unknown burdens incurred in its previous lives. The soul of Meister Eckhart may also have attained nirvana, and it causes no theoretical problem to Eastern religions that Meister Eckhart was governed by nirvana by way of a belief in God that is totally alien to them. It is a frequently cited saying in the East that there are many ways to nirvana. Nirvana, it would appear, also plays a mixed strategy.

Someone raised in a Western religion may find such a profane statement disturbing, even someone highly respectful of other religions, but such an attitude can be easily reconciled with an Eastern outlook on life, since there, belief itself is not so valuable. The only important thing is whether one becomes better in this life, gets closer to nirvana. Nirvana governs our soul willy-nilly. In this sense, our fate is determined. One can do battle against one's fate—and since one is imperfect, one does—but the game is not worth the candle, because the only result will be increased suffering in future lives.

One's soul may become better in many ways. The Eastern viewpoint does not preclude the existence of methods by which nirvana might also be reached that fall outside its meditative ways of cognition based on the belief in reincarnation. It is meet and right that souls born in Europe were born in Europe, because that is where it will be most suitable for them to move in the direction of nirvana. Péter Popper relates the following thoughts of an Indian doctor: "Family planning and the propagation of contraception are hopeless in India, because we do not restrict the *souls* who wish to be born, to *incarnate*, or to take the form of a body again.… In fact…we also accept those souls who should be born in Europe, but you do not accept them." The doctor also deduced from this that many Indian young people feel a spiritual connection with Western culture.

Rationality, the belief in the power of reason, has deeply infiltrated Western culture. Not only science, but religion, too, tries to express itself rationally, defining rationally its dogmas, dogmas that point beyond the power of reason. In accordance with the demands of Western thinking, everything that conforms to these dogmas is to be understood rationally. We have not specified the type of rationality, but this is not the important thing. Pure rationality, no matter what form it takes, means a kind of harmony, similar to that of pure mysticism. Perhaps it is this pure harmony that enables science to make discoveries with broad applicability despite its basically limited set of questions.

Science inserts formal logic between the consciousness that performs the act of cognition and that which is to be understood. This purely rational procedure is the essence of the scientific method, but it may also work as a kind of meditative technique of distancing. The most profound achievements of science may take one to a *state of mind* in which one can almost mystically live through the revealed unity of the universe. For example, one may learn about meaningful and stable equilibria arising through simple mixed strategies based on pure chance. In this state of mind one accepts the world as it is and perceives in it a deeply hidden harmony. There are many ways to nirvana, and some of them may involve one or another form of pure rationality.

References and Further Reading

1. Axelrod, Robert. *The Evolution of Cooperation*. Basic Books, 1984.

2. Bancroft, A. *Zen: Direct Pointing to Reality*. Thames and Hudson, 1979.

3. Bányai, É.I. and Hilgard, E.R. "A comparison of active-alert hypnotic induction with traditional relaxation induction." *Journal of Abnormal Psychology* 85 (1976), 218–224.

4. Barkow, J.H., Cosmides, C., and Tooby, J. *The Adapted Mind: Evolutionary Psychology and the Generation of Culture*. Oxford University Press, 1988.

5. Benjafield, J.G. *Cognition*. Prentice–Hall, 1992.

6. Bergson, H. *Creative Evolution*. Macmillan, 1964.

7. Black, M. *The Prevalence of Humbug*. Cornell University Press, 1983.

8. Bloom, F.E., Lazerson, A., and Hofstadter, L. *Brain, Mind, and Behavior*. Freeman, 1985.

9. Capra, Fritjof. *The Tao of Physics: An Exploration of the Parallels between Modern Physics and Eastern Mysticism*. Shambala, 1975.

10. Capra, Fritjof. *The Turning Point: Science, Society, and the Rising Culture*. Simon & Schuster, 1982.

11. Casti, John L. *Five Golden Rules: Great Theories of 20th Century Mathematics—And Why They Matter*. Wiley & Sons, 1996.

12. Cavalli-Sforza, L.L. and Feldman, M.W. *Cultural Transmission and Evolution. A Quantitative Approach*. Princeton University Press, 1981.

13. Cherniak, C. *Minimal Rationality*. MIT Press, 1986.

14. Cleary, T. *Rational Zen. The Mind of Dógen Zenji.* Shambala, 1992.

15. Colman, A.M. *Game Theory and Experimental Games.* Pergamon Press, 1982.

16. Colman, A.M. *Game Theory and Its Applications in Social and Biological Sciences.* Butterworth–Heineman, 1995.

17. Crick, F. and Koch, C. "The Problem of Consciousness," in *Brain and Mind, Readings from Scientific American.* Freeman, 1993.

18. Curtiss, S. *Genie.* Academic Press, 1977.

19. Damasio, Antonio R. *Descartes' Error: Emotion, Reason, and the Human Brain.* G.P. Putnam, 1994.

20. Davies, P. *The Mind of God.* Simon & Schuster, 1992.

21. Davis, Morton D. *Game Theory: A Nontechnical Introduction.* Basic Books, 1970.

22. Dawkins, Richard. *The Selfish Gene.* Oxford University Press, 1976.

23. Dawkins, Richard. *The Extended Phenotype: The Gene as the Unit of Selection.* Freeman, 1982.

24. Dennett, Daniel C. *Darwin's Dangerous Idea: Evolution and the Meanings of Life.* Simon & Schuster, 1995.

25. Donald, M. *Origins of the Modern Mind: The Stages in the Evolution of Culture and Cognition.* Harvard University Press, 1991.

26. Dupré, J. (ed.) *The Latest on the Best: Essays on Evolution and Optimality.* MIT Press, 1987.

27. Dupré, J. *The Disorder of Things: Metaphysical Foundations of the Disunity of Science.* Harvard University Press, 1993.

28. Eckhart, Johannes (Meister Eckhart). *Meister Eckhart, a Modern Translation.* Raymond Bernard Blakney, trans. Harper & Brothers, 1941.

29. Eigen, M. and Winkler, R. *Laws of the Game: How the Principles of Nature Govern Chance.* Random House, 1981.

30. Einstein, Albert. *Mein Weltbild.* Fischer, 1979. Translated as *Essays in Science.* Philosophical Library, 1955.

31. Einstein, Albert, Born, Hedwig, and Born, Max. *The Correspondence between Albert Einstein and Max and Hedwig Born: 1916–1955.* Walker & Company, 1971.

32. Enomiya-Lassalle, Hugo M. *Zen—Way to Enlightenment*. Taplinger, 1968.

33. Ewald, P.W. *Adaptation and Disease*. Oxford University Press, 1993.

34. Eysenck, M.W. *A Handbook of Cognitive Psychology*. Erlbaum, 1984.

35. Faust, D. *The Limits of Scientific Reasoning*. University of Minnesota Press, 1984.

36. Feynman, Richard P. *The Feynman Lectures on Physics*. Addison–Wesley, 1963–1965.

37. Feynman, Richard P. *The Character of Physical Law*. MIT Press, 1965.

38. Freud, Sigmund. *The Interpretation of Dreams*. Macmillan, 1915.

39. Freud, Sigmund. *A General Introduction to Psychoanalysis*. Boni and Liveright, 1920.

40. Glynn, I.M. "Consciousness and Time." *Nature* 348 (6301), 1990, 477–479.

41. Hadamard, J. *The Psychology of Invention in the Mathematical Field*. Dover 1945.

42. Harsányi, J.C. "Advances in Understanding Rational Behavior," in P.K. Moser, (ed.) *Rationality in Action*. Cambridge University Press, 1990.

43. Hawking, Stephen W. *A Brief History of Time: From the Big Bang to Black Holes*. Bantam Books, 1988.

44. Heisenberg, Werner. *Physics and Beyond: Encounters and Conversations*. Harper & Row, 1971.

45. Hilgard, E.R. *The Experience of Hypnosis*. Harcourt, Brace and World, 1968.

46. Hofstadter, Douglas R. *Gödel, Escher, Bach—An Eternal Golden Braid*. Basic Books, 1979.

47. Hofstadter, Douglas R. and Dennett, Daniel. *The Mind's I: Fantasies and Reflections on Self and Soul*. Basic Books, 1981.

48. Hofstadter, Douglas R. *Metamagical Themas: Questing for the Essence of Mind and Pattern*. Basic Books, 1985.

49. Hookway, C. (ed.) *Mind, Machine and Evolution*. Cambridge University Press, 1985.

50. Huizinga, Johan. *Homo Ludens: A Study of the Play-Element in Culture*. Beacon Press, 1955.

51. Jahn, R.G. (ed.) *The Role of Consciousness in the Physical World.* Westview Press, 1981.

52. Jahn, R.G. and Dunne, B.J. *Margins of Reality. The Role of Consciousness in the Physical World.* Harvest/HBJ, 1987.

53. Johnson, P.E. *Darwin on Trial.* InterVarsity Press, 1991.

54. Johnsson-Laird, P.M. *Mental Models.* Cambridge University Press, 1983.

55. Johnsson-Laird, P.M and Byrne, R.M.J. *Deduction.* Cambridge University Press, 1993.

56. Jones, R.H. "Rationality and Mysticism." *International Philosophical Quarterly* XXVII (1987), 263–279.

57. Jung, Carl G. *Memories, Dreams, Reflections.* Richard and Clara Winston, trans. Vintage Books, 1965.

58. Jung, Carl G. "Two Essays on Analytical Psychology" in *The Collected Works of C. G. Jung,* vol. 7. Random House, 1966.

59. Kahn, Hermann. *On Escalation: Metaphors and Scenarios.* Frederick A. Praeger, 1965.

60. Kant, Immanuel. *Foundations of the Metaphysics of Morals.* Bobbs–Merrill, 1969.

61. Kelley, H.H. and Thibaut, J.W. *Interpersonal Relations.* Wiley, 1978.

62. Komorita, S.S. and Parks, C.D. *Social Dilemmas.* Brown & Benchmark Social Psychology Series, 1994.

63. Kornai, J. *Anti-Equilibrium.* Közgazdasági és Jogi Kiadó, 1975.

64. Kostolany, Andre. *Kostolanys Börsenpsychologie: Vorlesungen am Kaffeehaustisch.* Düsseldorf, Econ, 1991.

65. Kuhn, Thomas. *The Structure of Scientific Revolutions.* University of Chicago Press, 1962.

66. Lecron, L.M. *Self Hypnotism.* Prentice–Hall, 1964.

67. Lederman, Leon. *The God Particle: If the Universe Is the Answer, What Is the Question?* Houghton Mifflin, 1993.

68. Leininger, W. "Escalation and cooperation in conflict situations: The dollar auction revisited." *Journal of Conflict Resolution* 33 (1989), 231–254.

69. Lewin, R. *Complexity.* Collier Books, Macmillan, 1992.

70. Ling, T. *A History of Religion: East and West.* Macmillan, 1968.

71. Lorenz, Konrad. *The Foundations of Ethology.* Springer-Verlag, 1981.

72. Luce, R.D. and Raiffa, H. *Games and Decisions.* John Wiley & Sons, 1957.

73. Lumsden, C.J. and Wilson, E.O. *Genes, Mind and Culture.* Harvard University Press, 1981.

74. Marshall, I.N. "Consciousness and Bose–Einstein Condensates." *New Ideas in Psychology* 7 (1989), 73–83.

75. Maynard Smith, J. "Group Selection." *Quarterly Review of Biology* 51 (1964), 277–283.

76. Maynard Smith, J. *Evolution and the Theory of Games.* Cambridge University Press, 1982.

77. Maynard Smith, J. *Evolutionary Genetics.* Cambridge University Press, 1989.

78. Mazlish, B. *The Fourth Discontinuity.* Yale University Press, 1993.

79. McCarthy, K.A. "Indeterminacy and consciousness in the creative process: What quantum physics has to offer." *Creativity Research Journal* 6 (1993), 201–219.

80. Mealy, L. "The sociobiology of sociopathy: An integrated evolutionary model." *Behavioral and Brain Sciences* 18 (1995), 523–541.

81. Mérő, L. "The 'least number wins' game on a large sample." *Proc. 10th International Conference of the PME.* London, 1986, 457–462.

82. Mérő, L. *Ways of Thinking: The Limits of Rational Thought and Artificial Intelligence.* trans. Anna C. Gösi-Greguss, ed. Viktor Meszaros. World Scientific Publishing Co., 1990.

83. Milinski, M. "Tit for Tat in sticklebacks and the evolution of cooperation." *Nature* 325 (1987), 433–435.

84. Monod, Jacques. *Chance and Necessity: An Essay on the Natural Philosophy of Modern Biology.* Knopf, 1971.

85. Moser, P.K. (ed.) *Rationality in Action.* Cambridge University Press, 1992.

86. Neumann, John von and Morgenstern, Oskar. *Theory of Games and Economic Behavior.* Princeton University Press, 1947.

87. Neumann, John von. *Mathematical Foundations of Quantum Mechanics.* Princeton University Press, 1955.

88. Osheron, D.N. and Smith, E.E. *An Invitation to Cognitive Science,* vol. 3, "Thinking," MIT Press, 1990.

89. Peat, F.D. *Synchronicity: The Bridge Between Matter and Mind.* Bantam Books, 1987.

90. Penrose, Roger. *The Emperor's New Mind: Concerning Computers, Minds, and the Laws of Physics.* Oxford University Press, 1989.

91. Penrose, Roger. *Shadows of the Mind: A Search for the Missing Science of Consciousness.* Oxford University Press, 1994.

92. Piaget, Jean. *Play, Dream, and Imitation in Childhood.* Norton, 1962.

93. Poincaré, Henri. *The Foundations of Science: Science and Hypothesis; The Value of Science; Science and Method.* Science Press, 1913.

94. Poundstone, W. *Prisoner's Dilemma.* Doubleday, 1992.

95. Pruitt, D.G. "Reward structure and cooperation: The decomposed prisoner's dilemma game." *Journal of Personality and Social Psychology* 7 (1977), 21–27.

96. Rao, K.R. "Consciousness, awareness, and first-person perspective." *Behavioral and Brain Sciences* 16 (1993), 415–416.

97. Rapoport, A. (ed.) *Game Theory as a Theory of Conflict Resolution.* D. Riedel Publishing Company, 1974.

98. Rapoport, A. *Experimental Studies of Interactive Decisions.* Kluwer, 1990.

99. Rapoport, A. and Chammah, A.M. *Prisoner's Dilemma.* University of Michigan Press, 1965.

100. Russell, Bertrand. *Mysticism and Logic.* Doubleday, 1957.

101. Samuelson, Paul A. *Economics.* McGraw-Hill, 1970.

102. Schelling, T.C. *The Strategy of Conflict.* Harvard University Press, 1960.

103. Schelling, T.C. *Choice and Consequence.* Harvard University Press, 1984.

104. Schwartz, B. *Psychology of Learning and Behavior.* W.W. Norton & Company, 1989.

105. Searle, John R. "Minds, Brains and Programs." *Behavioral and Brain Sciences* 3 (1980), 417–457.

106. Searle, John R. "Consciousness, explanatory inversion, and cognitive science." *Behavioral and Brain Sciences* 13 (1990), 585–642.

107. Searle, John R. *The Rediscovery of the Mind.* MIT Press, 1992.

108. Shanon, B. "Cognitive psychology and modern physics: Some analogies." *European Journal of Cognitive Psychology* 3 (1990), 201–234.

109. Shubik, Martin. "The dollar auction game: A paradox in non-cooperative behavior and escalation." *Journal of Conflict Resolution* 15 (1971), 109–111.

110. Shubik, Martin. *Game Theory in the Social Sciences.* MIT Press, 1982.

111. Sigmund, Karl. *Game Dynamics, Mixed Strategies and Gradient Systems.* Laxenburg, Austria: International Institute for Applied Systems Analysis, 1987.

112. Sigmund, Karl. *Games of Life: Explorations in Ecology, Evolution, and Behavior.* Oxford University Press, 1993.

113. Simon, H.A. "A mechanism for Social Selection and successful Altruism." *Science* 250 (1990), 1665–1668.

114. Simon, H.A. *Reason in Human Affairs.* Stanford University Press. 1983.

115. Simonyi, Karoly. *Kulturgeschichte der Physik von den Anfängen bis 1990.* H. Deutsch, 1995.

116. Slobodkin, L.B. *Simplicity & Complexity in the Games of Intellect.* Harvard University Press, 1992.

117. Smith, Adam. *An Inquiry into the Nature and Causes of the Wealth of Nations.* Modern Library, 1937.

118. Smullyan, Raymond. *The Tao is Silent.* Harper, 1957.

119. Smullyan, Raymond. *5000 B.C.* St. Martin's Press, 1983.

120. Sternberg, R.J. *Handbook of Human Intelligence.* Cambridge University Press, 1982.

121. Stewart, I. *Does God Play Dice?* Blackwell, 1989.

122. Suzuki, Daisetz T. *Mysticism: Christian and Buddhist; The Eastern and Western Way.* Collier, 1962.

123. Suzuki, Daisetz T., Fromm, Erich, and de Martino, Richard. *Zen Buddhism & Psychoanalysis.* Harper, 1960.

124. Székely, Gabor J. *Paradoxes in Probability Theory and Mathematical Statistics.* Reidel, 1986.

125. Talbot, M. *Mysticism and the New Physics.* Bantam Books, 1981.

126. Teger, A.I. *Too Much Invested to Quit.* Pergamon Press, 1980.

127. Thomas, L.C. *Game Theory and Applications.* Ellis Horwood, 1986.

128. Thorndike, Edward L. *Animal Intelligence: Experimental Studies.* Macmillan, 1911.

129. Trivers, R. *Social Evolution.* Benjamin Cummings, 1985.

130. Ulam, S. *Adventures of a Mathematician.* Scribner, 1976.

131. Varela, F.J., Thompson, E., and Rosch, E. *The Embodied Mind.* MIT Press, 1993.

132. Watts, Alan. *This is It, and Other Essays on Zen and Spiritual Experience.* Vintage Books, 1973.

133. Watzlawick, P., Weakland, John H., and Fisch, Richard. *Change: Principles of Problem Formation and Problem Resolution.* W.W. Norton, 1974.

134. Weber, Max. *The Theory of Social and Economic Organization.* Oxford University Press, 1947.

135. Weizenhoffer, A.M. and Hilgard, E.R. *Stanford Hypnotic Susceptibility Scale, Forms A and B.* Consulting Psychologists Press, 1959.

136. Wigner, E.P. "The unreasonable effectiveness of mathematics." *Comm. Pure and Applied Mathematics* 13 (1960), 1–14.

137. Williams, J.D. *The Compleat Strategist.* McGraw Hill, 1966.

138. Wilson, D.S. and Sober, E. "Reintroducing group selection to the human behavioral sciences." *The Behavioral and Brain Sciences* 17 (1994), 585–654.

139. Wooldridge, D. *The Mechanical Man—The Physical Basis of Intelligent Life.* McGraw–Hill, 1968.

140. Zajonc, R.B. "Feeling and thinking: Preferences need no inferences." *American Psychologist* 35 (1980), 151–175.

141. Zohar, D. *The Quantum Self.* Harper Collins, 1991.

142. Zohar, D. and Marshall, I. *The Quantum Society.* William Morrow and Company, 1994.

Index